# LATIN AMERICA IN GRAPHS

## Two Decades of Economic Trends

## 1971-1991

Published by the Inter-American Development Bank
Distributed by The Johns Hopkins University Press

Washington, D.C.
1992

**LATIN AMERICA IN GRAPHS**

Two Decades of Economic Trends (1971-1991)

© Copyright 1992 by the Inter-American Development Bank.

Inter-American Development Bank
1300 New York Avenue, N.W.
Washington, D.C. 20577

Distributed by
The Johns Hopkins University Press
701 West 40th Street
Baltimore, MD 21211

ISBN: 0-940602-57-1

## INTRODUCTION TO LATIN AMERICA IN GRAPHS, 1971-1991

This document has been prepared by Statistics and Quantitative Analysis in the Development Studies Sub-Department of the Economic and Social Development Department. It contains a selection of graphs for each of the Developing Member Countries of the Inter-American Development Bank and for Latin America, as an aggregate of them. The objective is to give a visualization of trends in important economic variables over the period 1971-1991. The use of standardized variables and graphs across countries makes the comparison among them possible as well as with the aggregates for Latin America.

The graphs have been grouped into four major sections covering the following topics:

1. **National Accounts:**
   GDP per capita
   GDP and Population growth rates
   GDP Structure by Major Sectors of Origin, 1991
   GDP Growth by Major Sectors of Origin
   Degree of Openness of the Economy
   Investment and Domestic Saving Rates

2. **Balance of Payments:**
   Current Account Balance as a Percent of GDP
   Current Account components as a Percent of GDP
   Exports and Imports of Goods FOB
   Private and Public Capital Flows
   Net Direct Investment as a Percent of GDP
   Change in International Reserves

3. **External Trade:**
   Structure of Exports of Goods in 1970, 1980 and 1990
   Structure of Imports of Goods in 1970, 1980 and 1990

4. **External Debt:**
   Disbursed External Debt as a Percent of GDP
   Structure of Disbursed External Debt, End 1991
   Multilateral Debt as a Percent of Disbursed External Debt
   Interest Payments as a Percent of Exports
   Implicit Interest Rate Versus LIBOR
   Debt and Accumulated Current Deficits

General comments are given on the graphs at the beginning of each section and notes containing country exceptions are attached at the end of them.

All of the data are taken directly from the Bank's Economic and Social Data Base (ESDB). Many of the series are published in Basic Socio-Economic Data, the Statistical Profiles of the Country Chapters and the Statistical Appendix of *Economic and Social Progress in Latin America, 1992 Report*, though for a shorter time period.

The graphs have been designed by Jacques Anas who also developed the program and procedures used to automate their production. Statistical and data base programming support was given to the project by Gabriela Aviles, Fernando Quevedo, Ivo Maric and Ivan Guerra. The project was co-ordinated by Michael McPeak.

# TABLE OF CONTENTS

### National Accounts                                                    Page

    General Comments .................................................................. 3
    Latin America Graphs............................................................... 5
    Country Graphs .................................................................. 7-31
    Country Notes ........................................................................ 32

### Balance of Payments

    General Comments ................................................................ 35
    Latin America Graphs............................................................. 37
    Country Graphs ................................................................ 39-63
    Country Notes ........................................................................ 64

### External Trade

    General Comments ................................................................ 67
    Latin America Graphs............................................................. 69
    Country Graphs ................................................................ 71-95
    Country Notes ........................................................................ 96

### External Debt

    General Comments ................................................................ 99
    Latin America Graphs............................................................ 101
    Country Graphs .............................................................. 103-127
    Country Notes ...................................................................... 128

# NATIONAL ACCOUNTS

## NATIONAL ACCOUNTS

All graphs are produced using data in 1988 US Dollars calculated by Statistics and Quantitative Analysis from official member country data. Population estimates are from CELADE and the United Nations Population Division.

### GDP PER CAPITA

Each country level of Gross Domestic Product (GDP) per capita is compared with the average level for Latin America (the aggregate of 25 IDB Developing Member Countries).

### GDP AND POPULATION GROWTH RATES

The graph shows when the economic growth rate of the country is sufficient or not to cover the population growth rate.

### GDP STRUCTURE BY MAJOR SECTORS, 1991

The structure by Major Sectors is shown for the last year available (1991) using the system of prices of 1988. This avoids any price distortion in inter-country comparisons of structure.

The structure is estimated at market prices except for Argentina, Barbados, Brazil, Guyana, Honduras and Suriname where it is estimated at factor cost. *Agriculture* includes Agriculture and Livestock Production, Forestry and Logging and Fishing. *Industry* includes Manufacturing, Construction and Electricity, Gas and Water: it would normally include Mining and Quarrying as well, but given the sector's importance in some countries, it is shown separately in this graph. *Services* includes Wholesale and Retail Trade, Transport and Communications, Financial, Government and Other Services.

### GDP GROWTH BY MAJOR SECTORS, 1971=100

The Major Sectors are the same as in the graph on the Structure by Major Sectors except that Mining and Quarrying is included in *Industry*.

### DEGREE OF OPENNESS OF THE ECONOMY

The ratios are Exports and Imports of Goods and Non-factor Services to GDP. The long-term upwards movement of the Export ratio in many countries of the region is indicative of the efforts made to foster participation in the international trade of Goods and Non-factor Services. Since the ratios are calculated in real terms, the graph indicates the changing openness in volume terms.

### INVESTMENT AND DOMESTIC SAVING RATES

Investment refers to Gross Fixed Capital Formation plus Change in Stocks. In most cases, the difference between Domestic Saving and Investment is, if not already negative, insufficient to cover interest to be paid abroad, even after considering grants received. In these cases, the country would need to use External Saving.

# LATIN AMERICA
## National Accounts in 1988 US Dollars

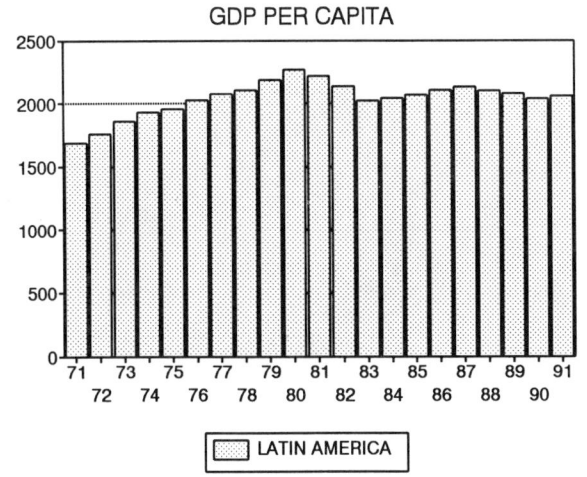

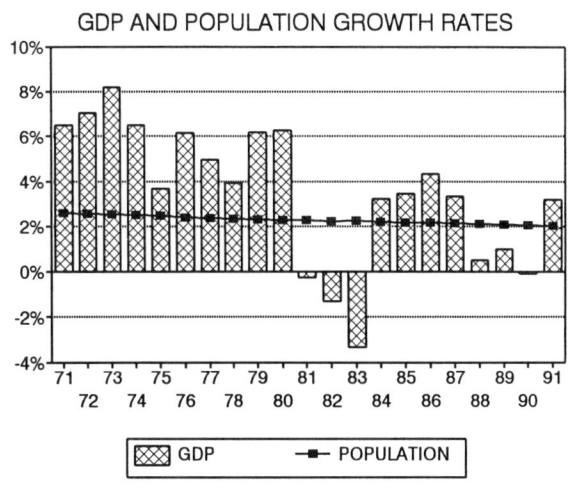

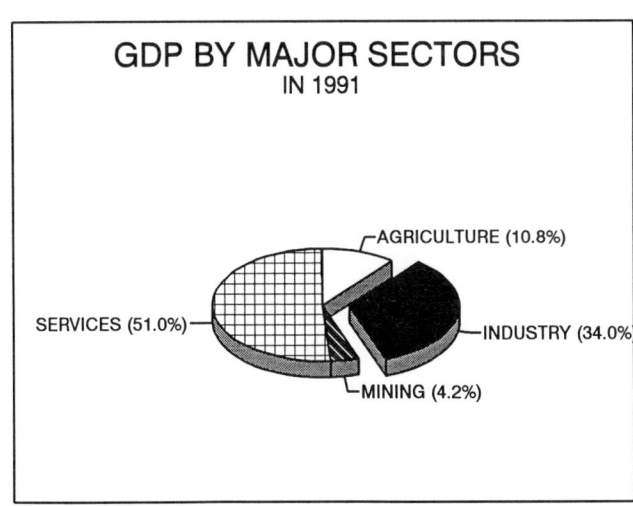

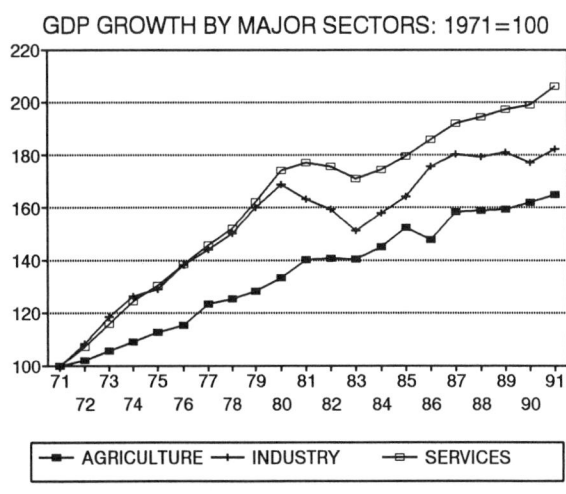

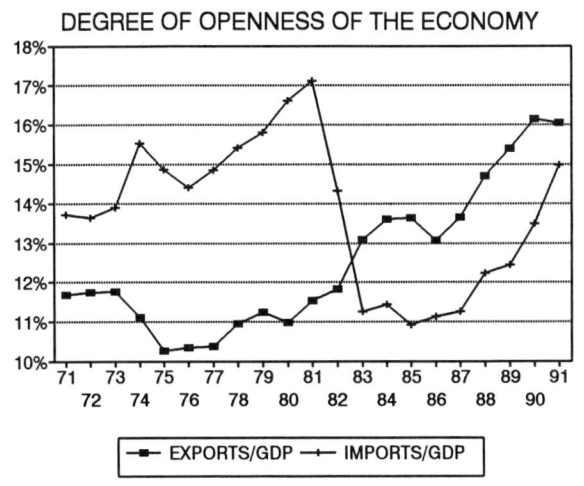

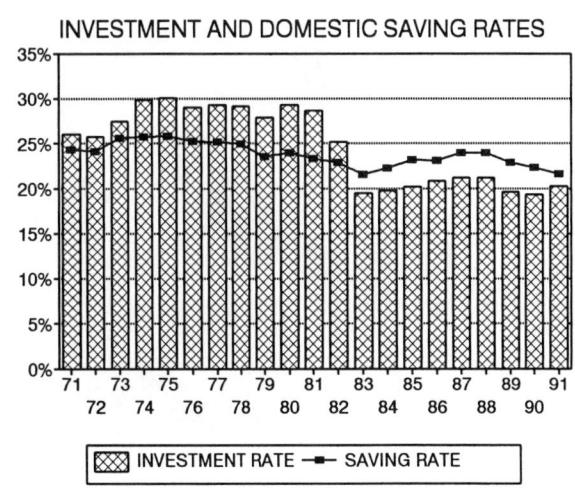

Statistics and Quantitative Analysis/IDB

# ARGENTINA
## National Accounts in 1988 US Dollars

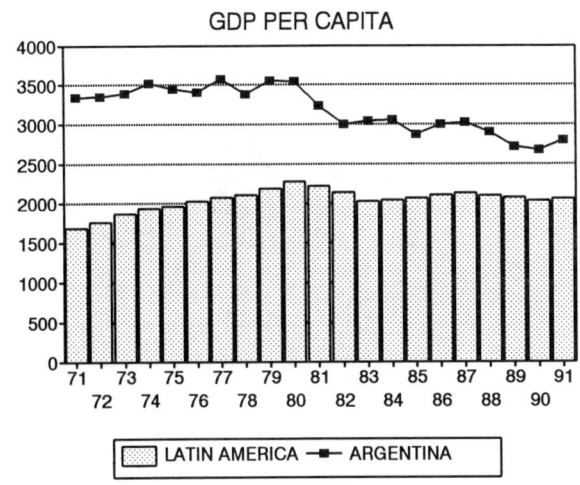

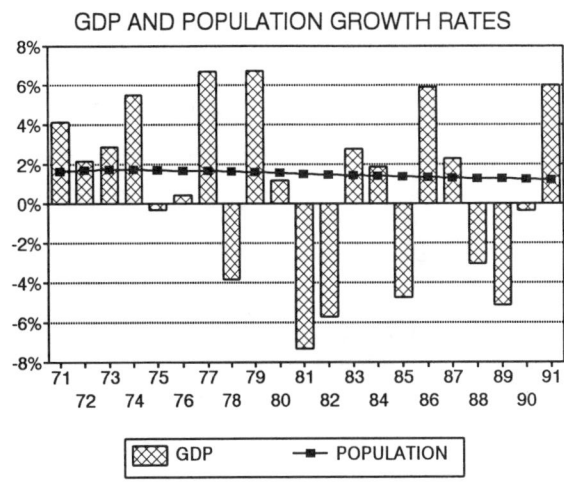

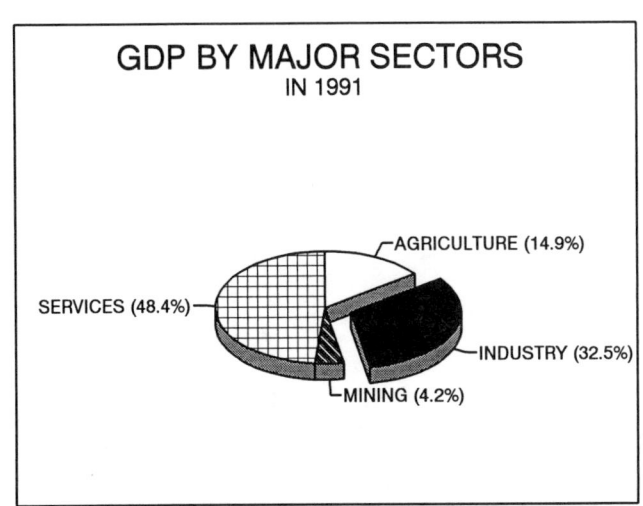

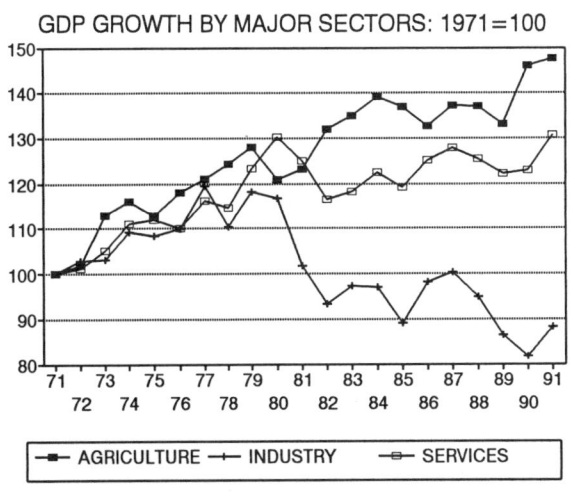

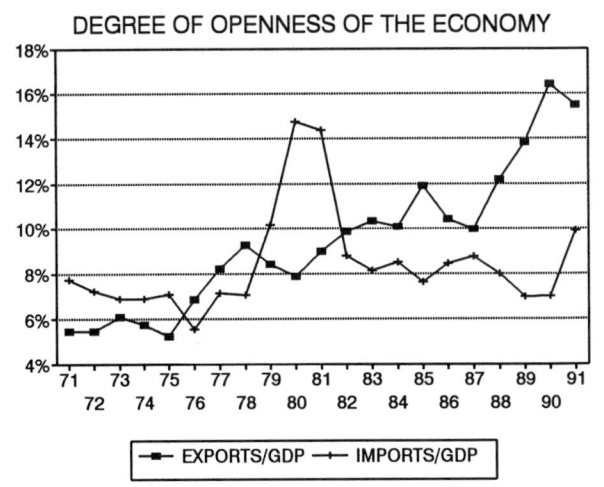

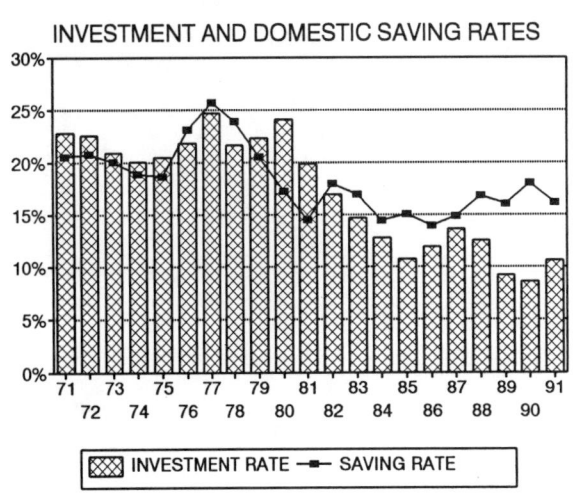

Statistics and Quantitative Analysis/IDB

# BAHAMAS
## National Accounts in 1988 US Dollars

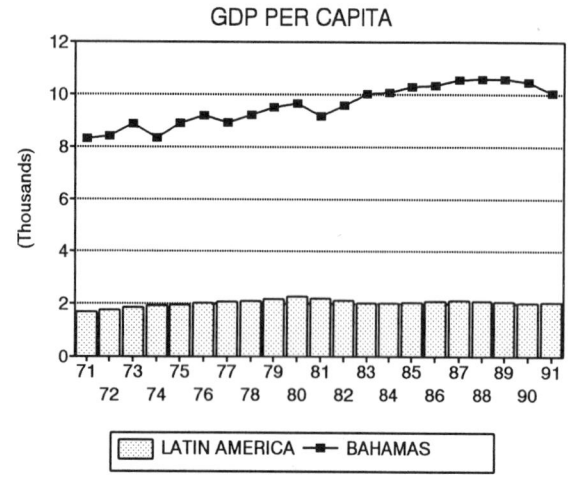
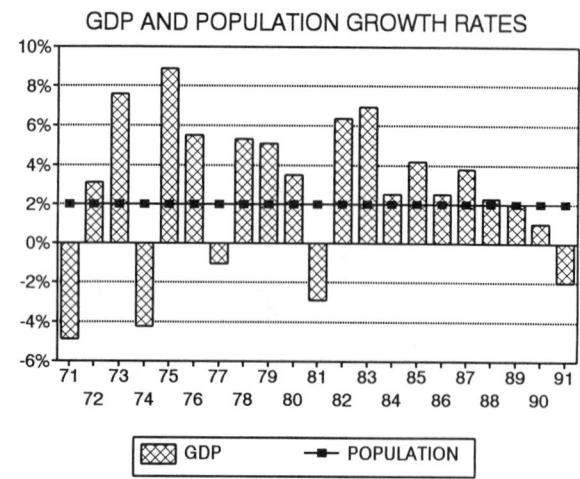

# BARBADOS
## National Accounts in 1988 US Dollars

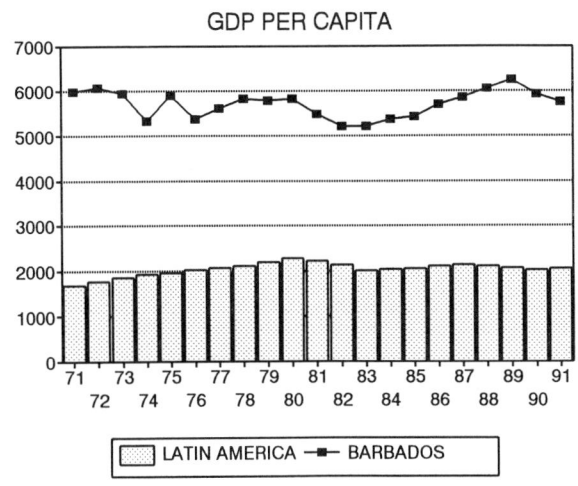

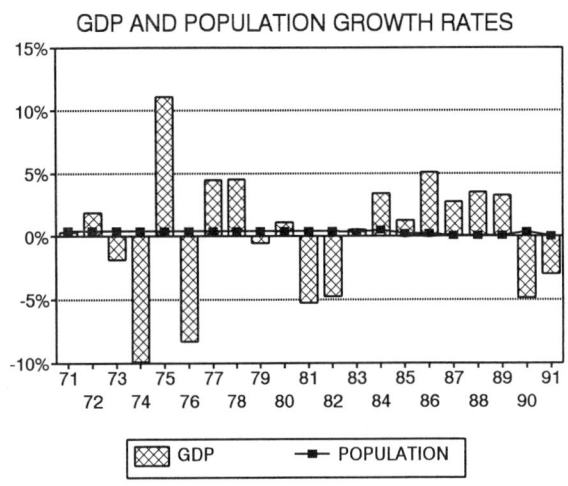

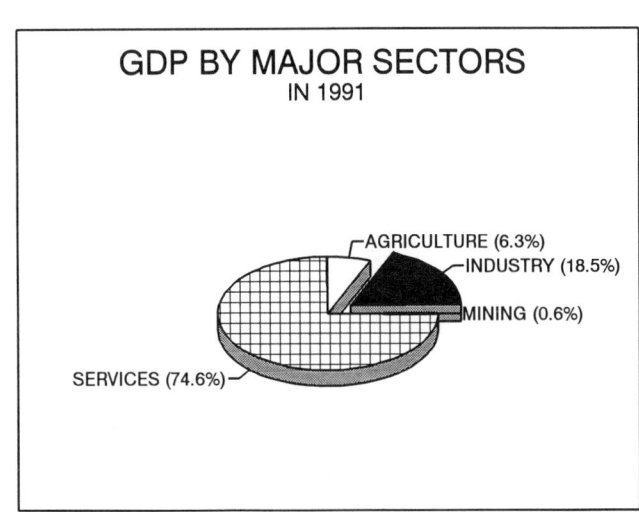

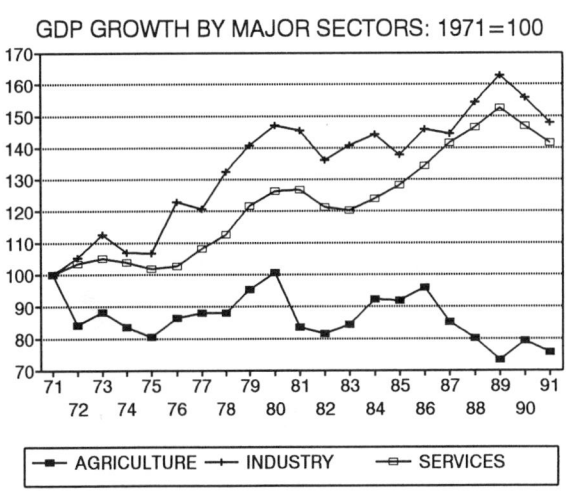

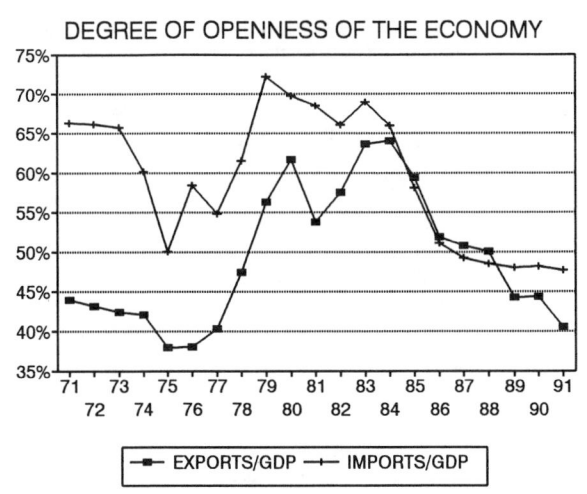

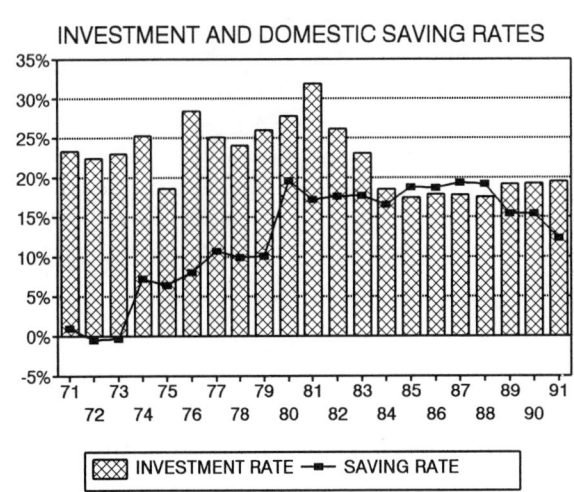

Statistics and Quantitative Analysis/IDB

# BOLIVIA
## National Accounts in 1988 US Dollars

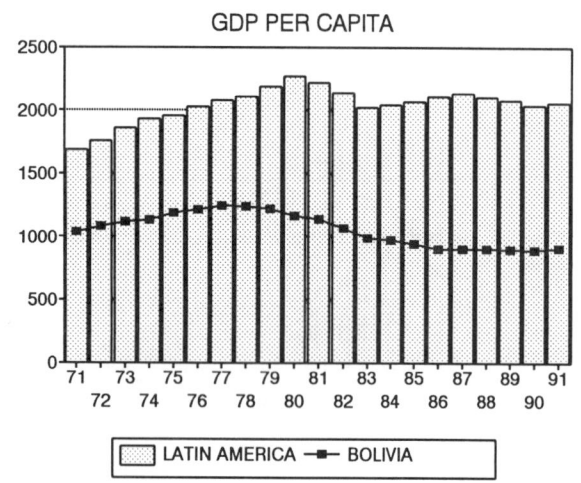

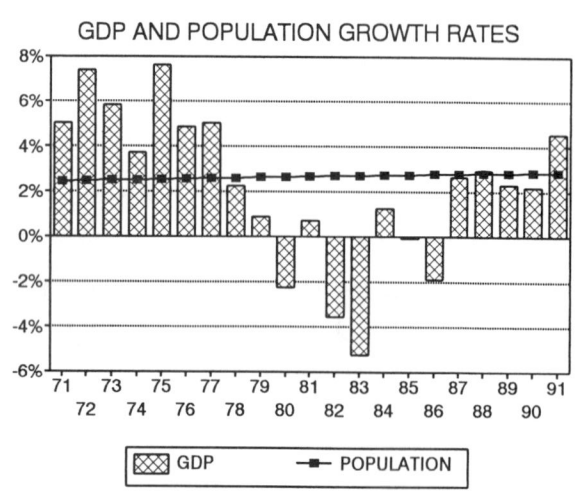

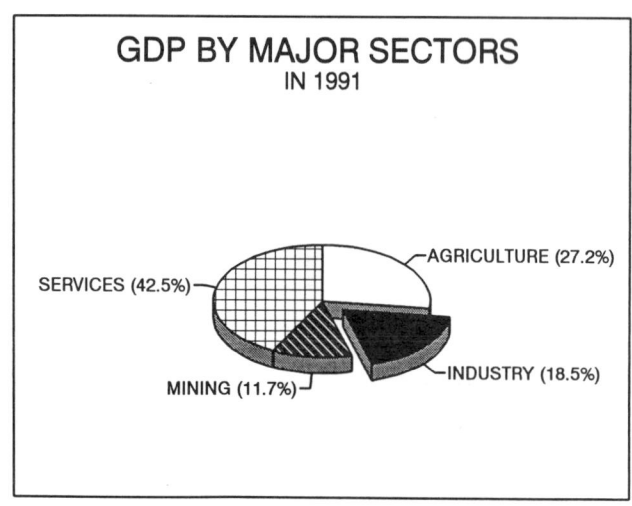

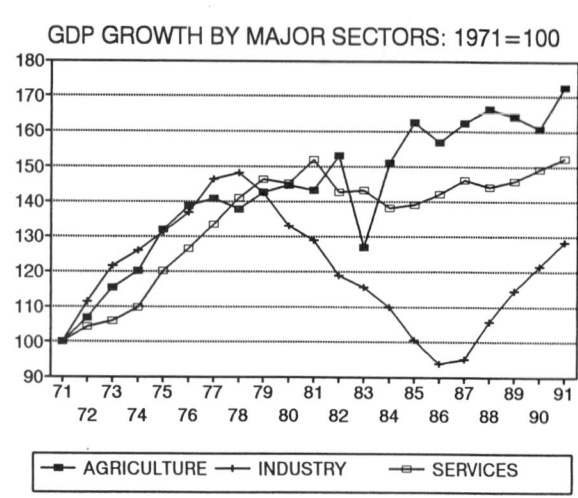

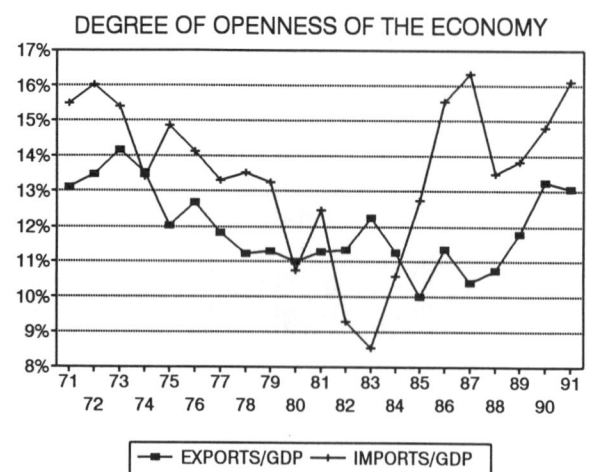

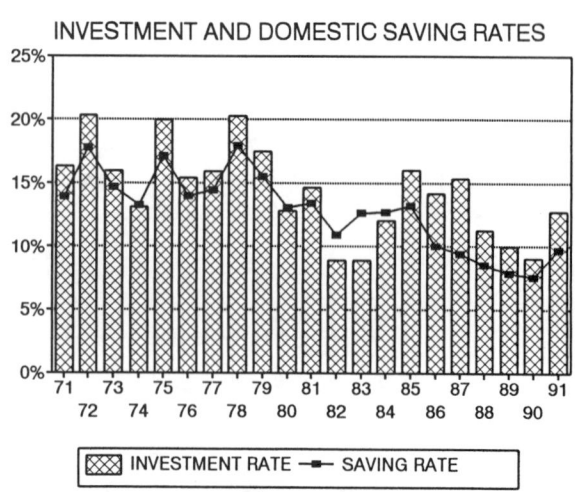

Statistics and Quantitative Analysis/IDB

# BRAZIL
## National Accounts in 1988 US Dollars

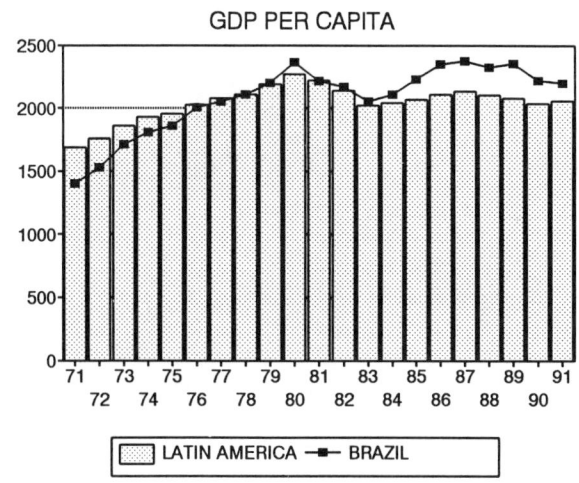

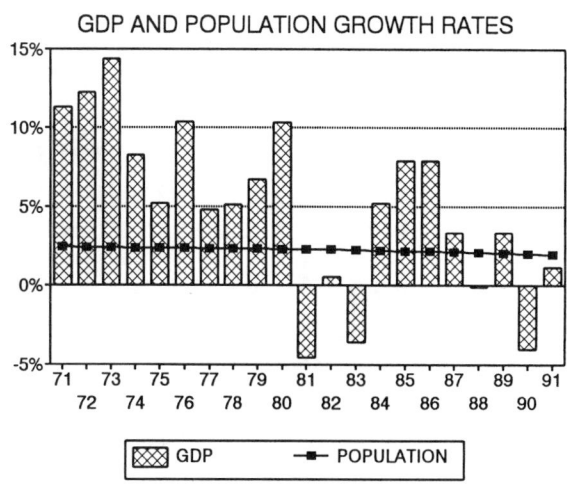

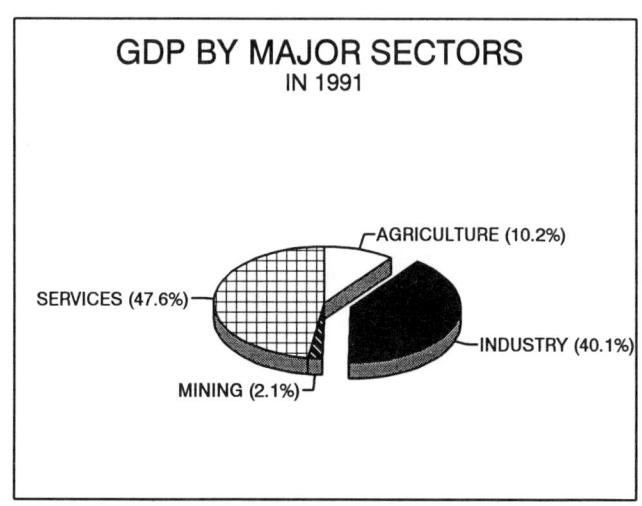

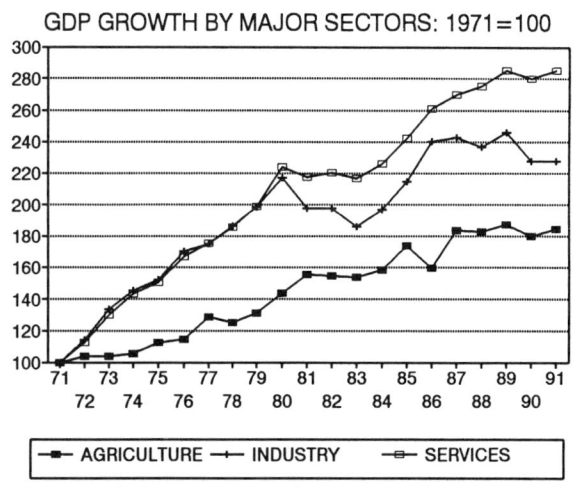

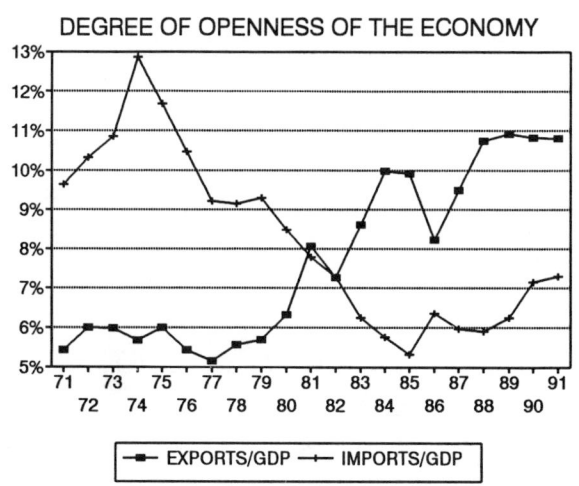

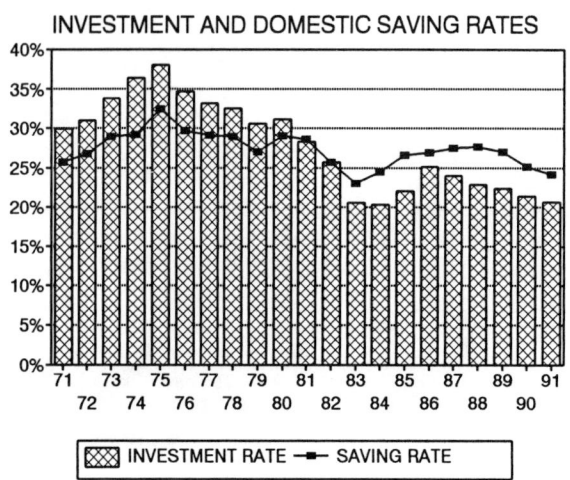

Statistics and Quantitative Analysis/IDB

# CHILE
## National Accounts in 1988 US Dollars

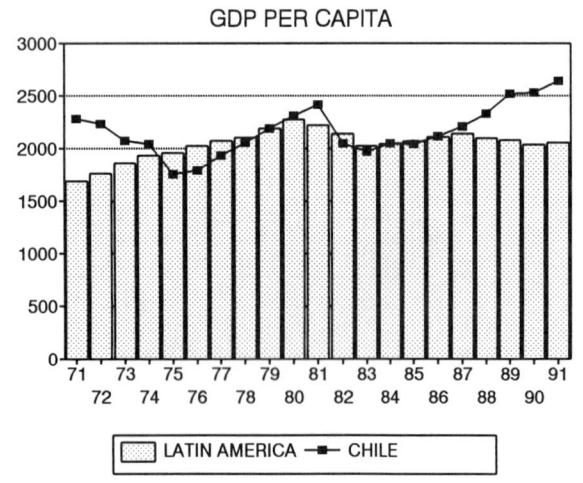

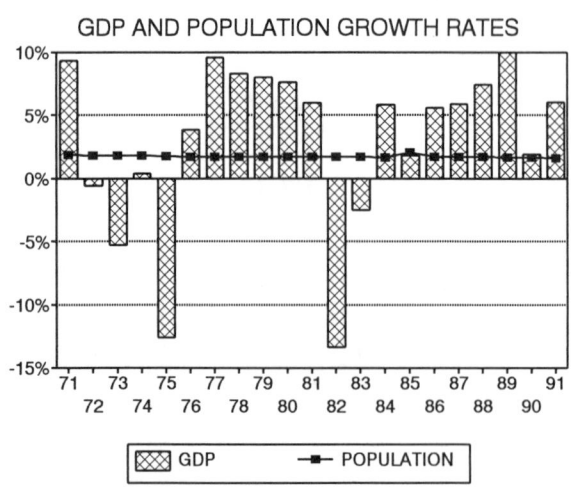

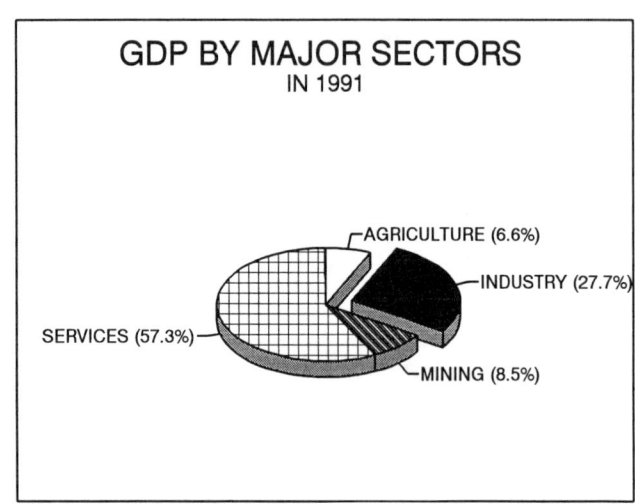

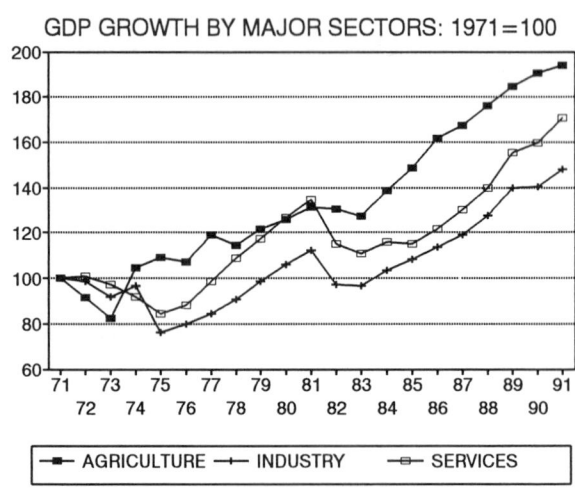

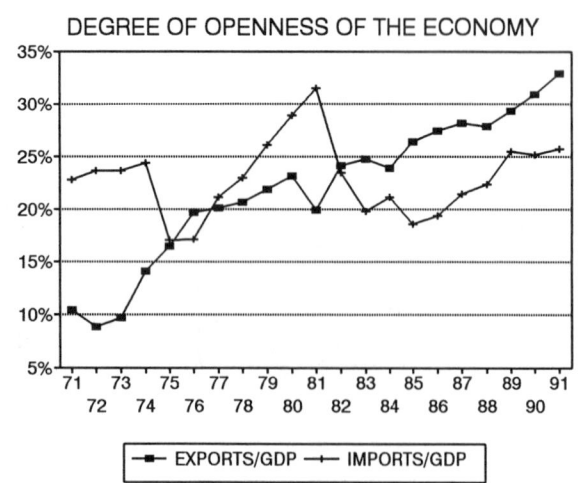

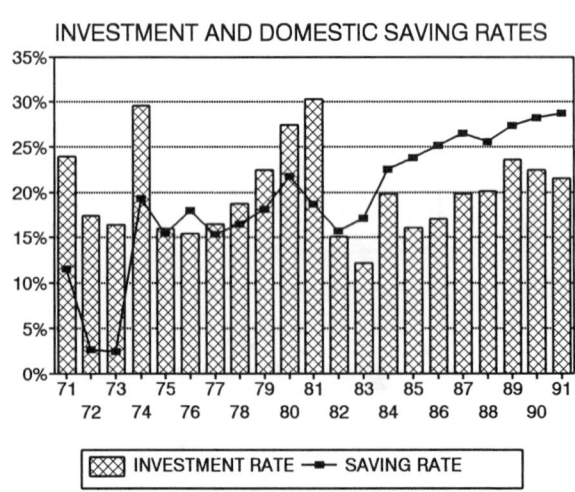

Statistics and Quantitative Analysis/IDB

# COLOMBIA
## National Accounts in 1988 US Dollars

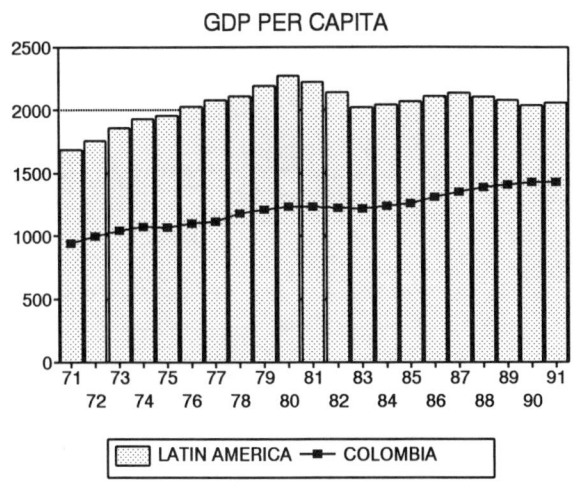

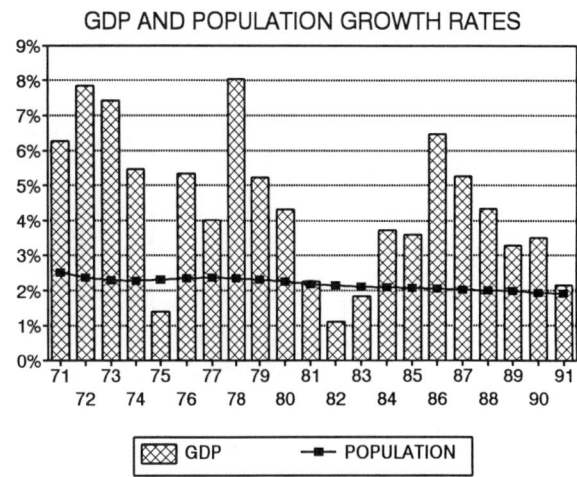

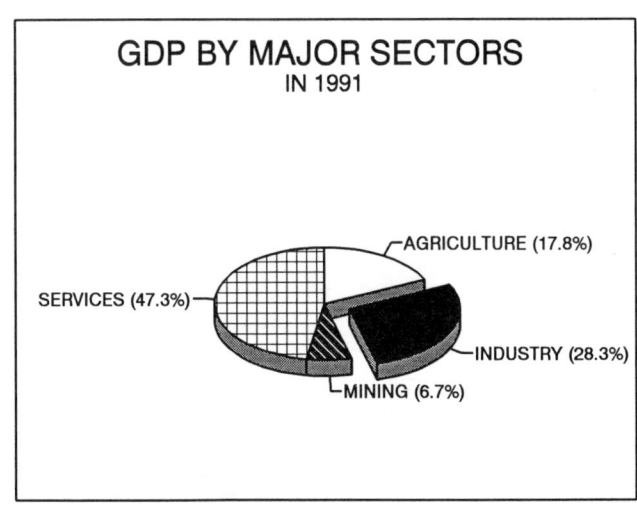

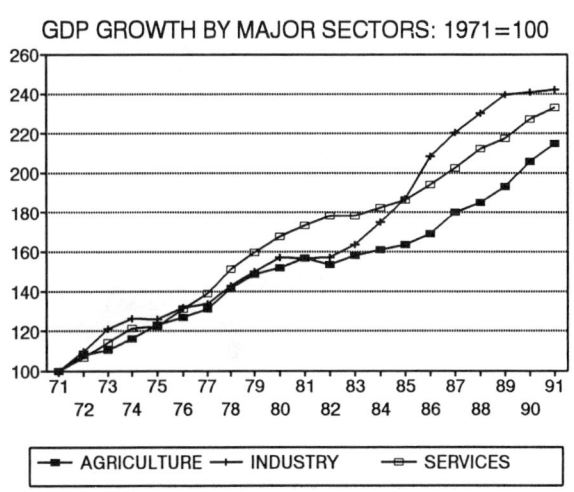

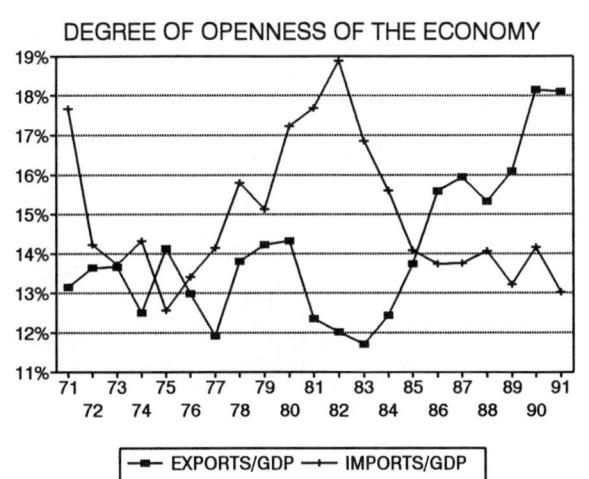

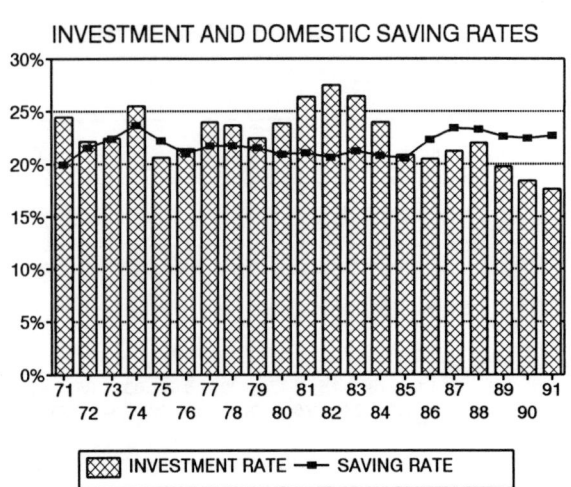

Statistics and Quantitative Analysis/IDB

# COSTA RICA
## National Accounts in 1988 US Dollars

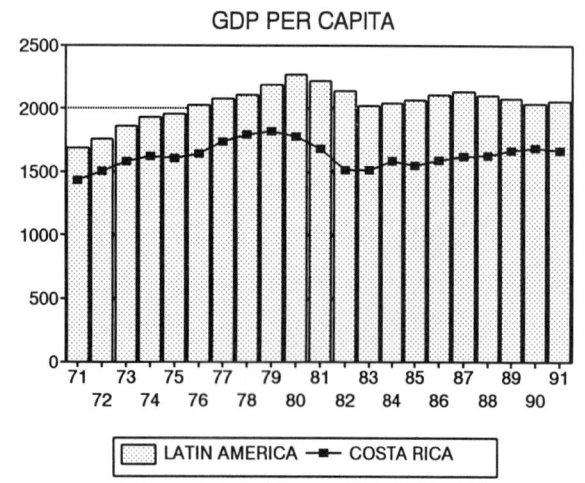

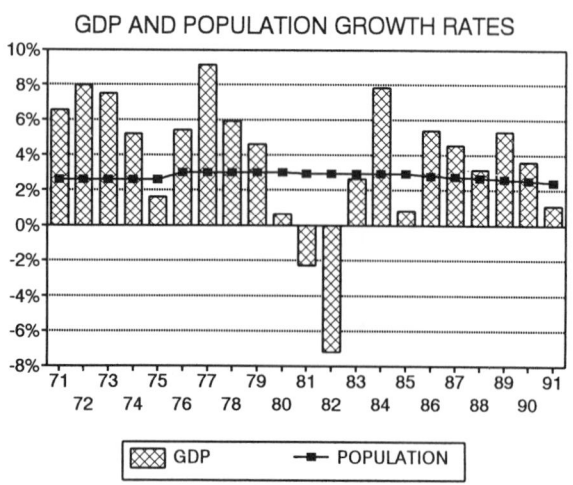

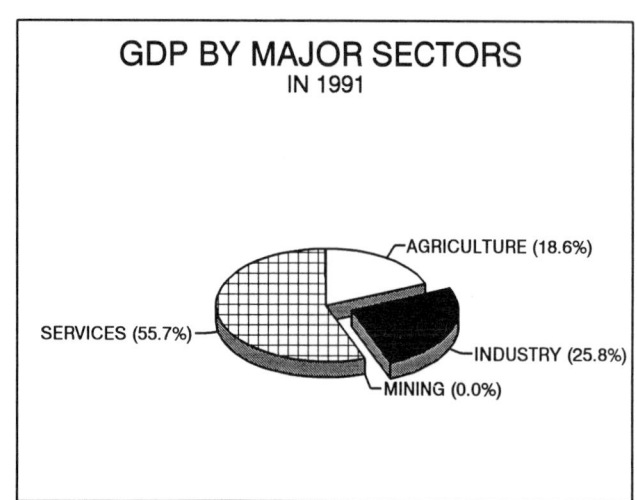

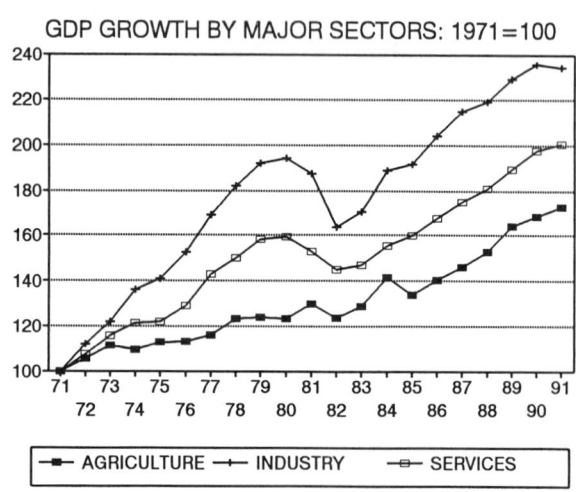

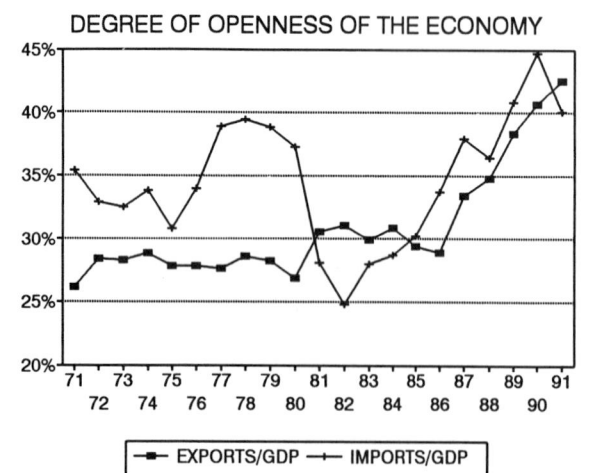

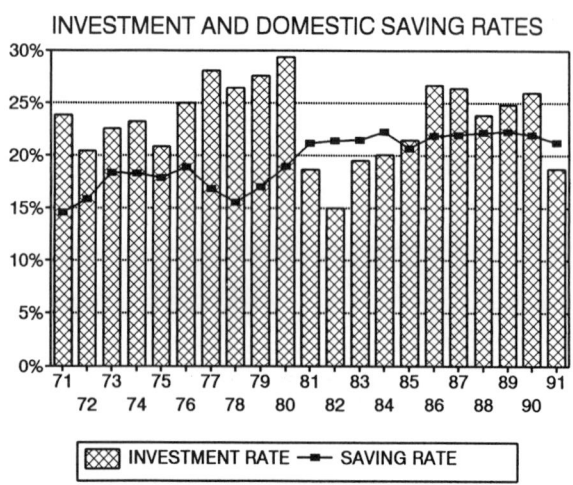

# DOMINICAN REPUBLIC
## National Accounts in 1988 US Dollars

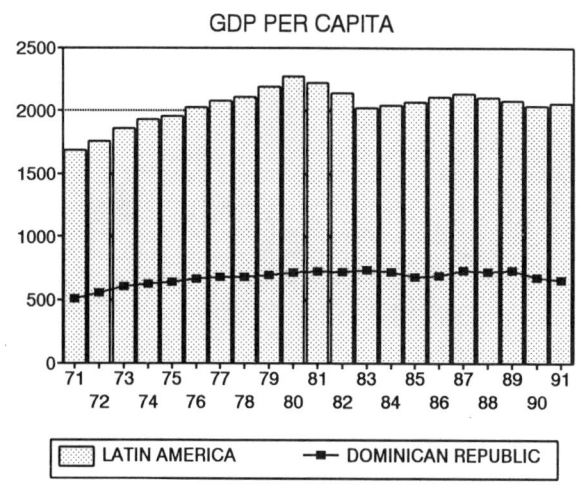

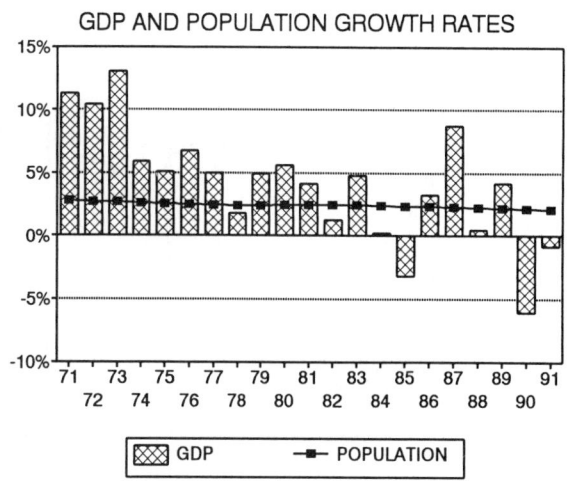

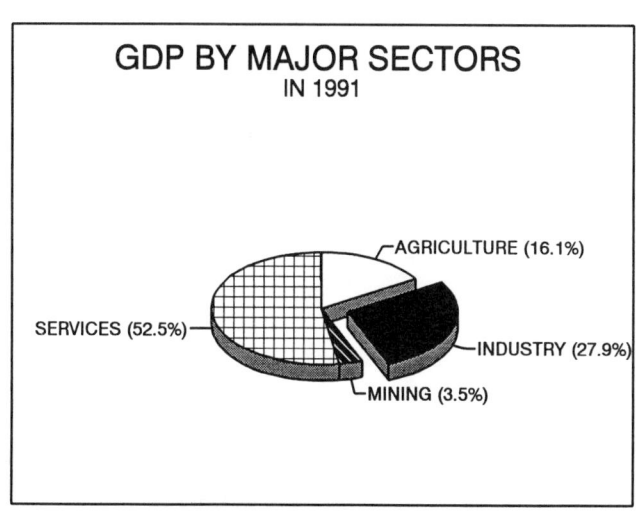

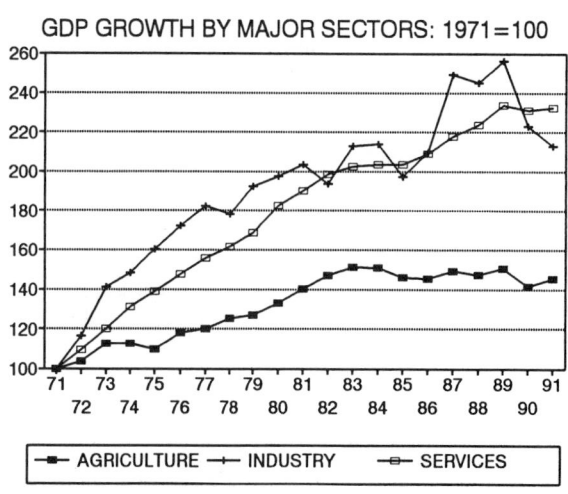

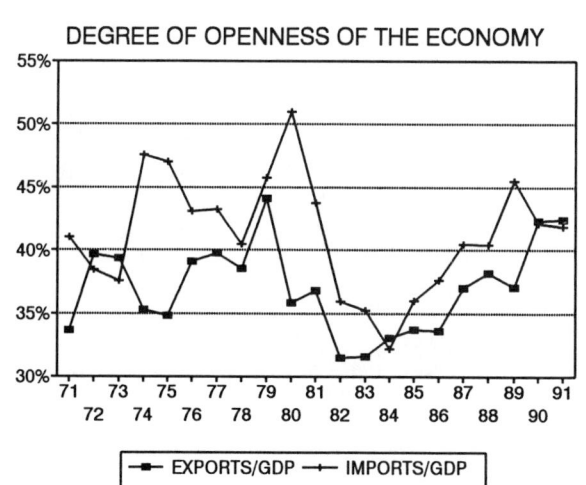

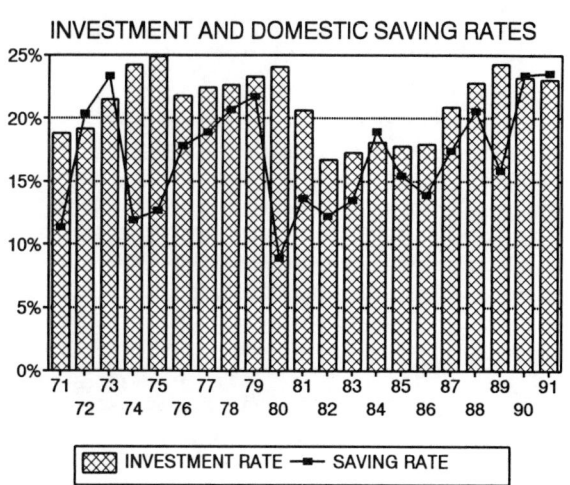

Statistics and Quantitative Analysis/IDB

# ECUADOR
## National Accounts in 1988 US Dollars

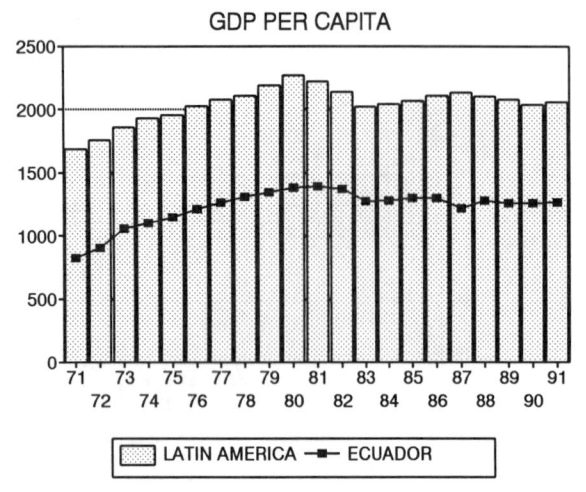

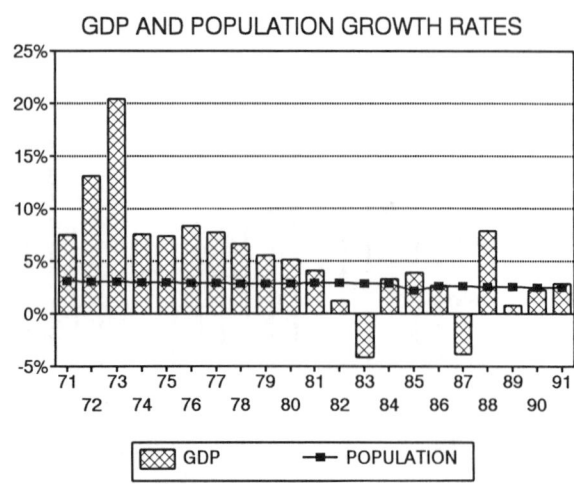

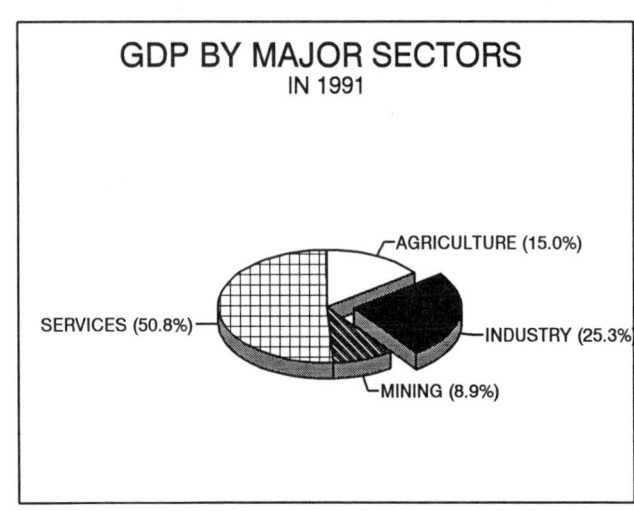

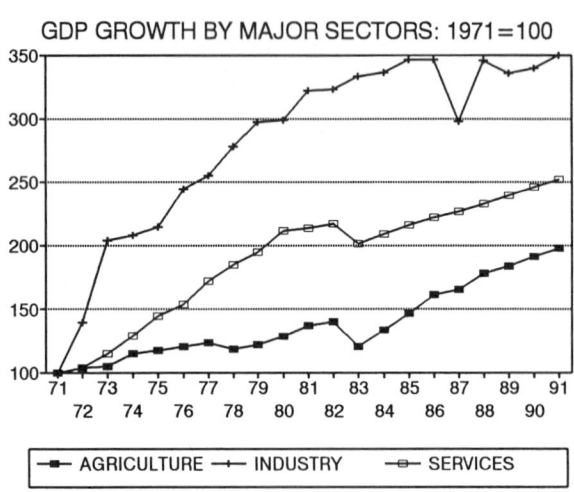

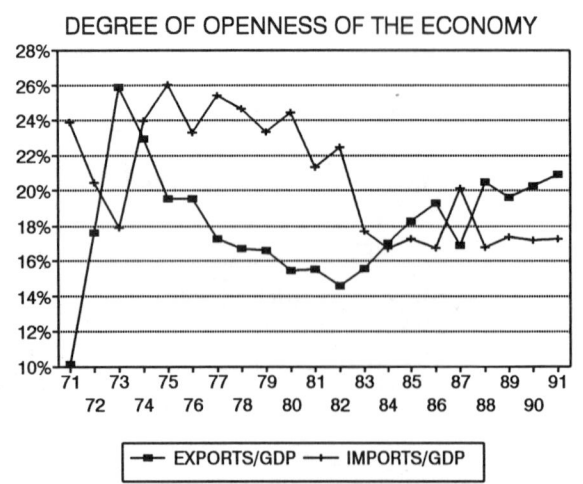

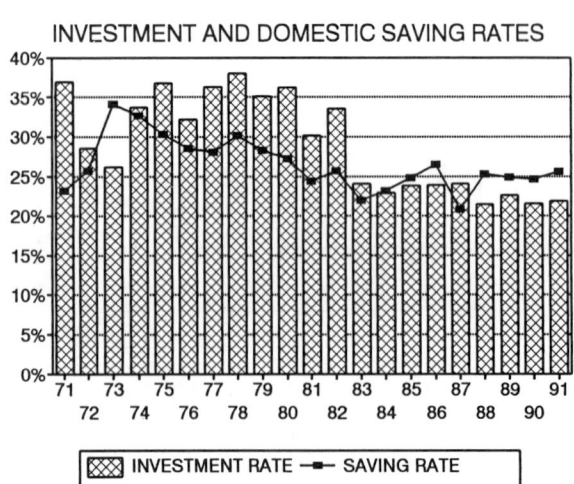

Statistics and Quantitative Analysis/IDB

# EL SALVADOR
## National Accounts in 1988 US Dollars

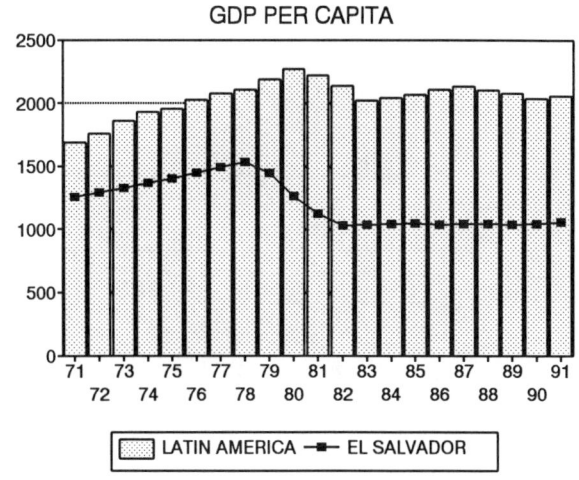

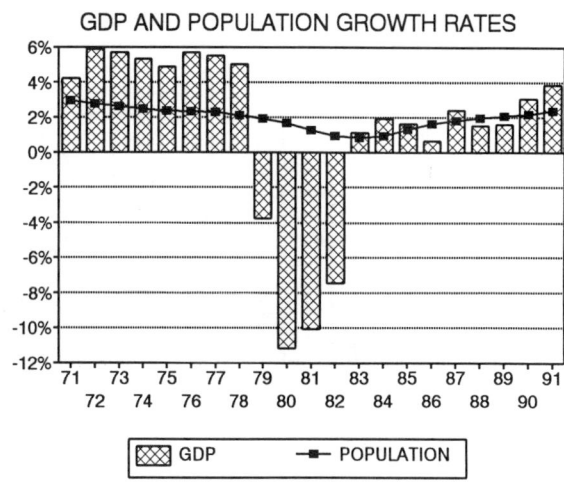

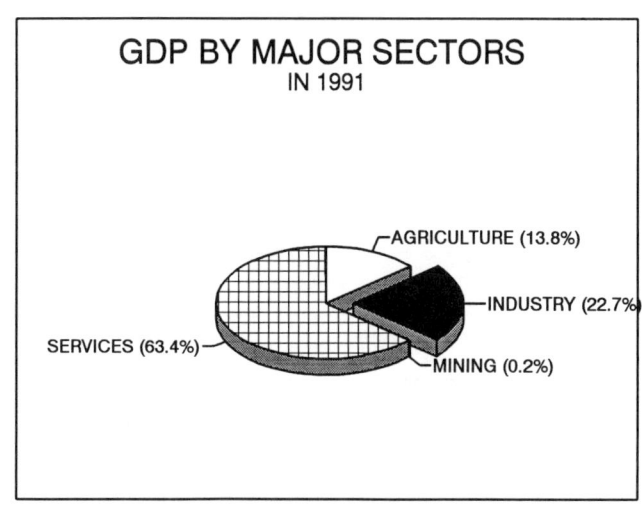

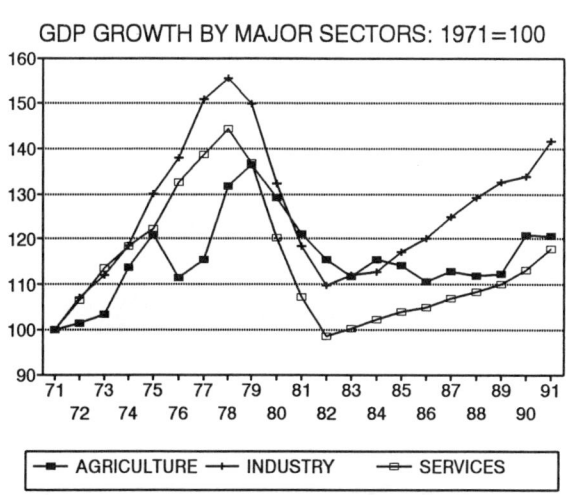

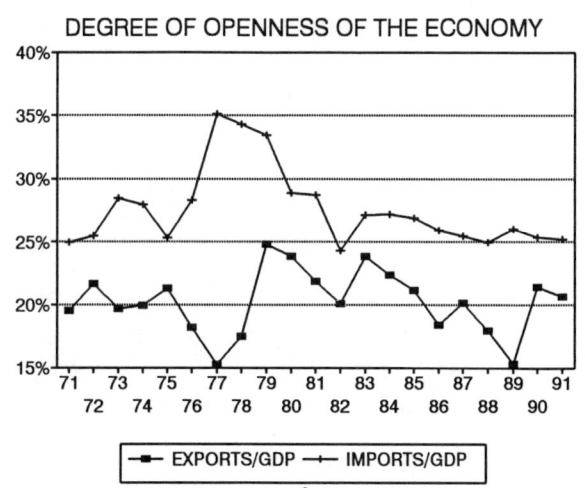

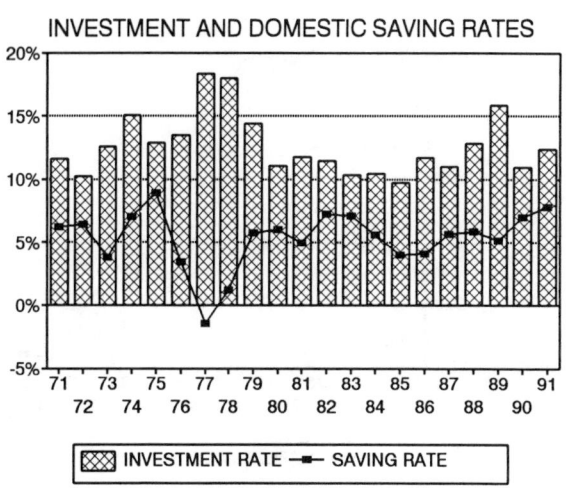

Statistics and Quantitative Analysis/IDB

# GUATEMALA
## National Accounts in 1988 US Dollars

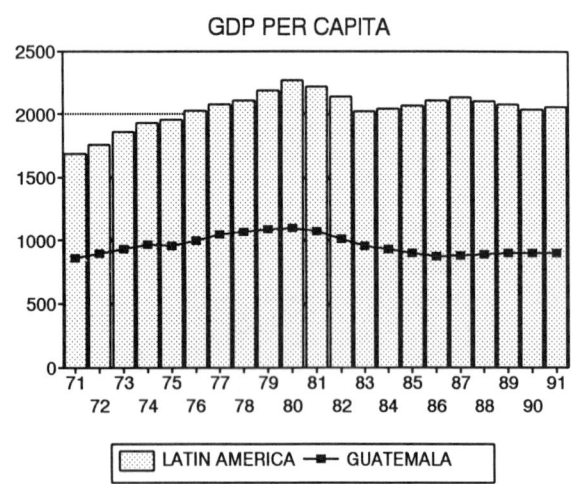

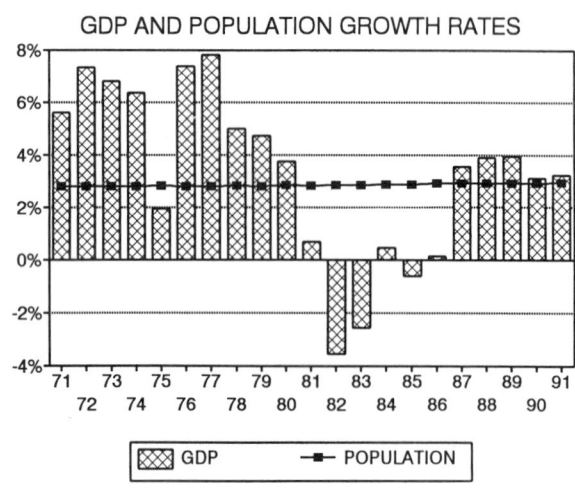

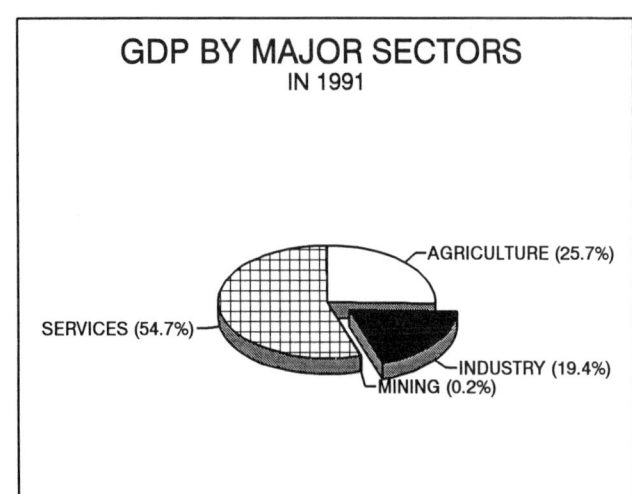

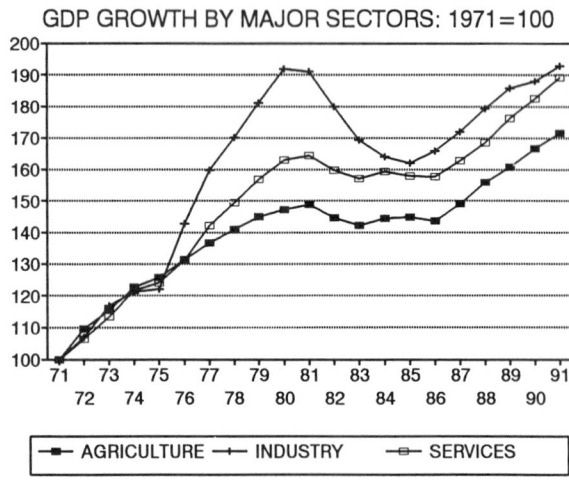

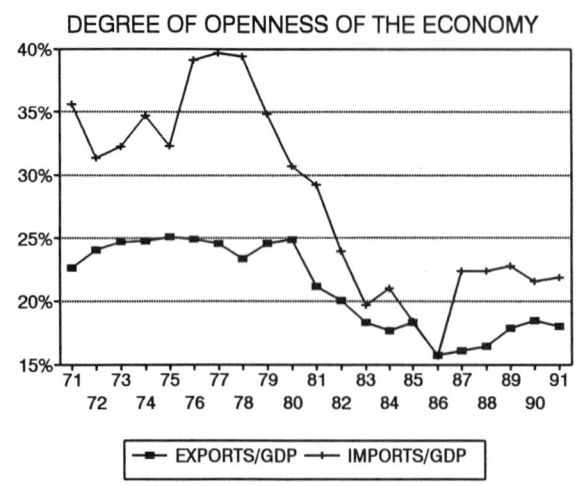

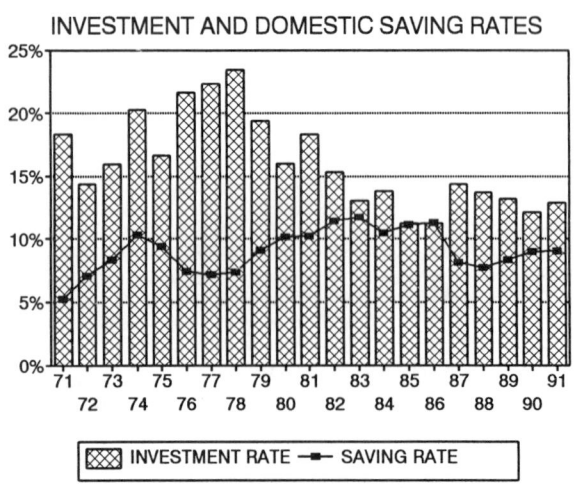

Statistics and Quantitative Analysis/IDB

# GUYANA
## National Accounts in 1988 US Dollars

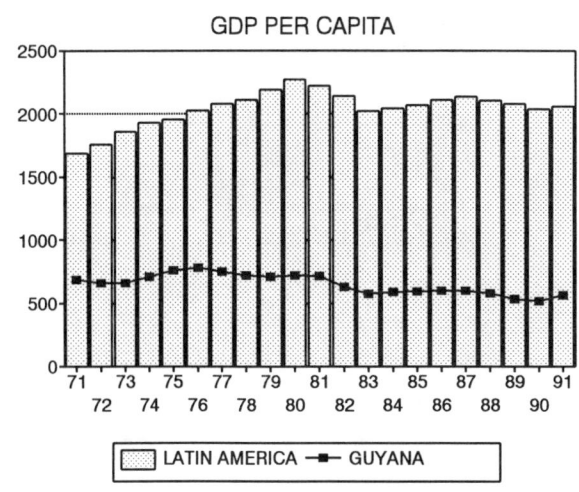

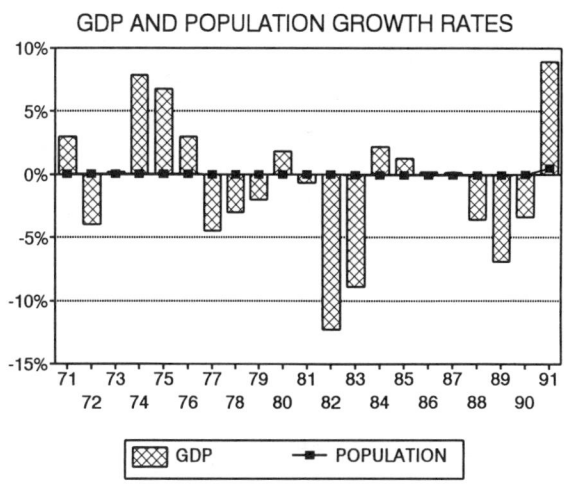

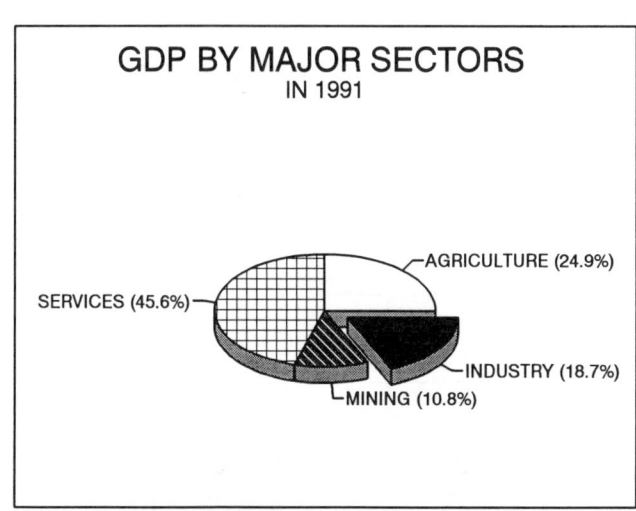

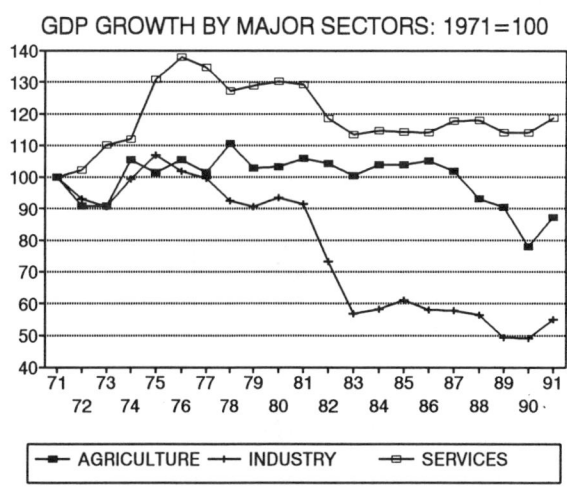

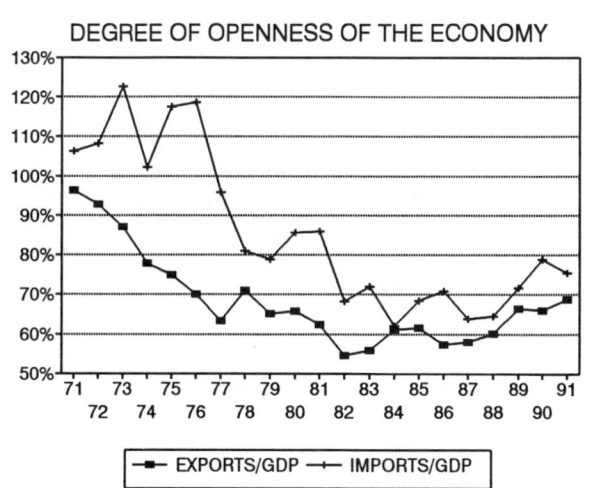

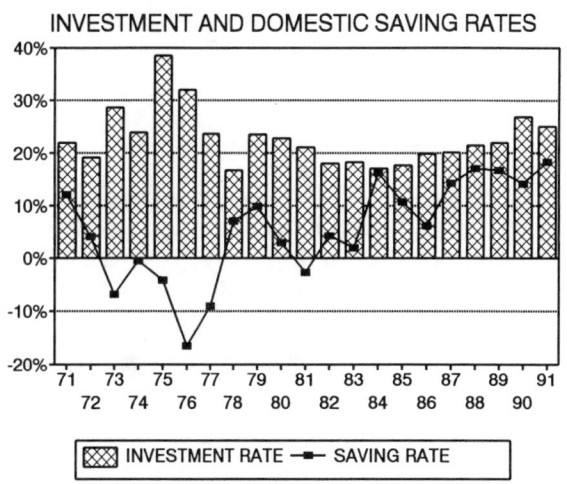

Statistics and Quantitative Analysis/IDB

# HAITI
## National Accounts in 1988 US Dollars

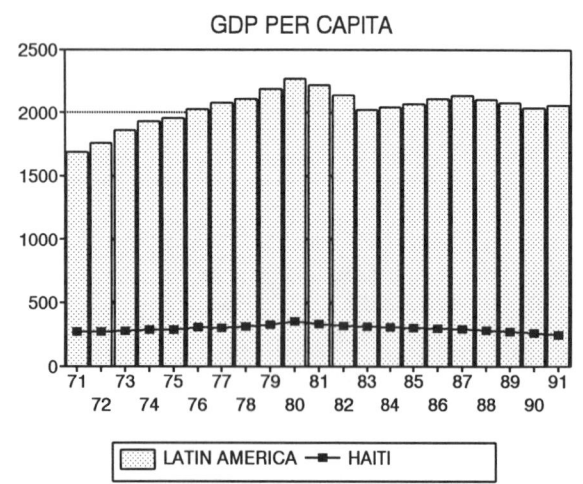
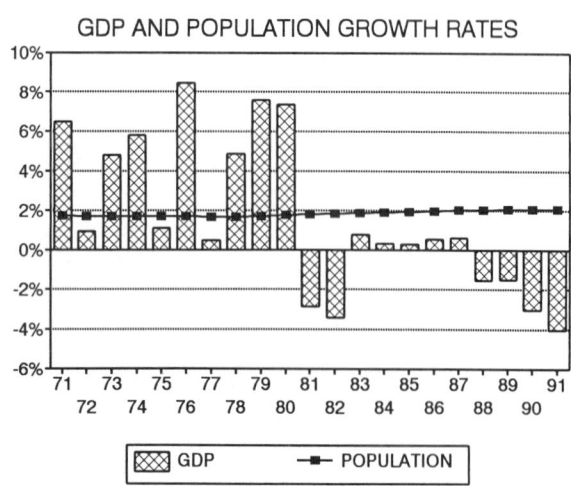
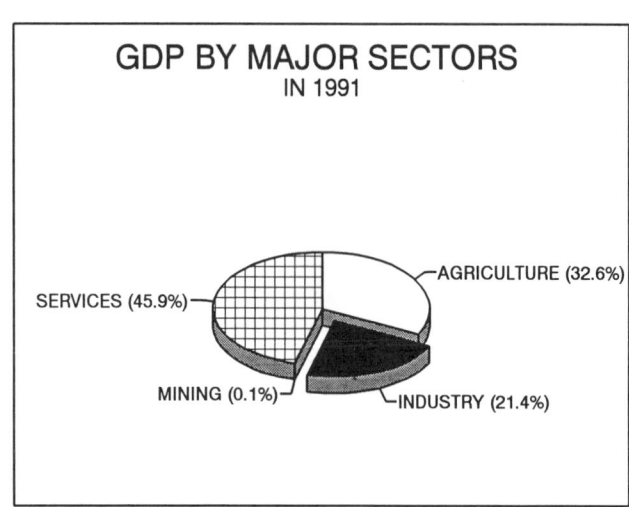
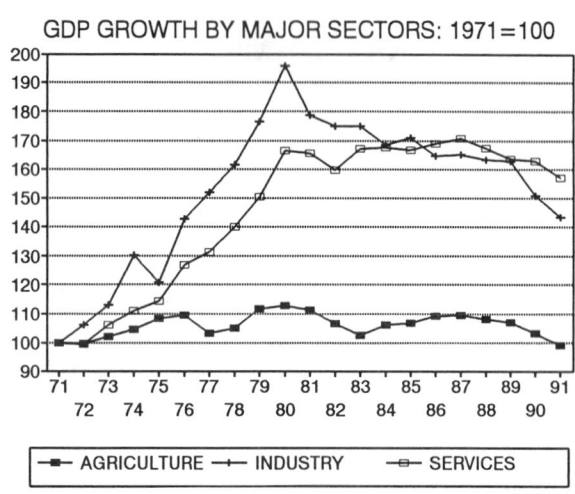
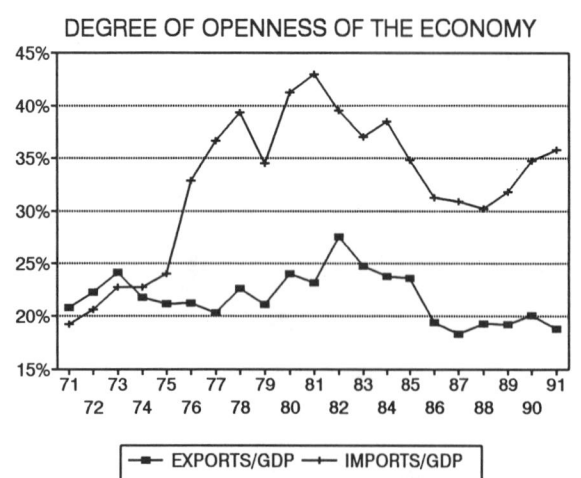
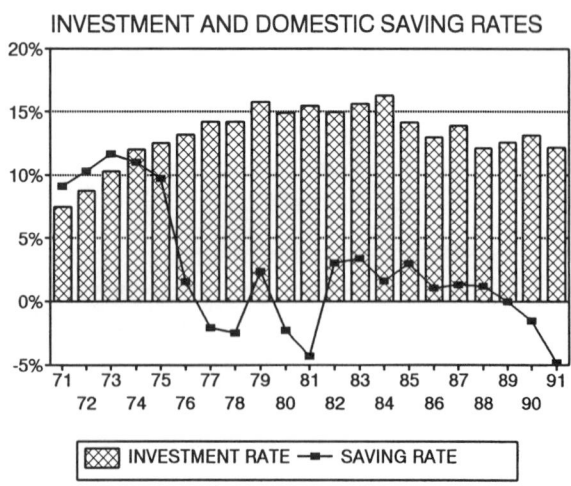

Statistics and Quantitative Analysis/IDB

# HONDURAS
## National Accounts in 1988 US Dollars

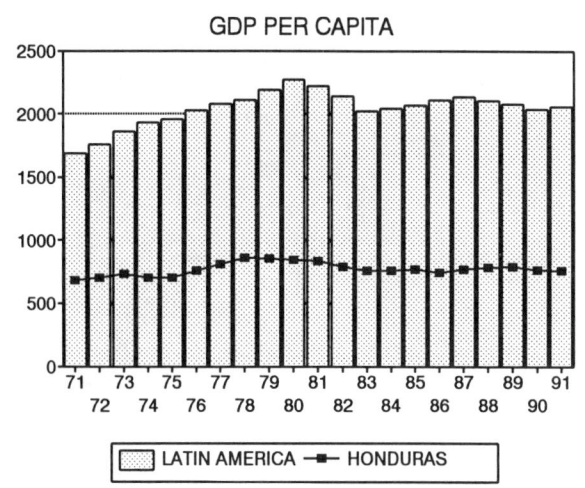

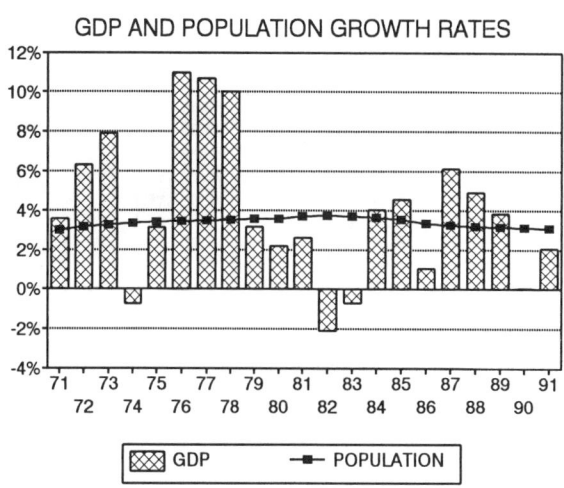

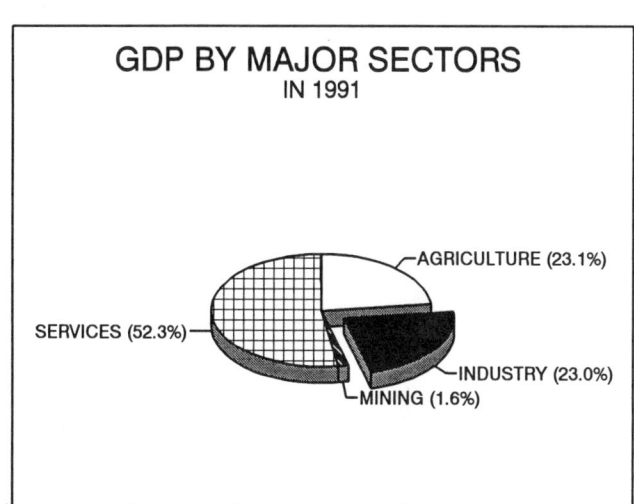

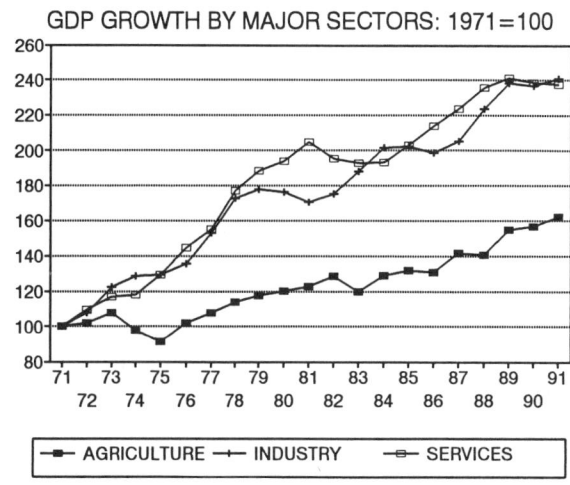

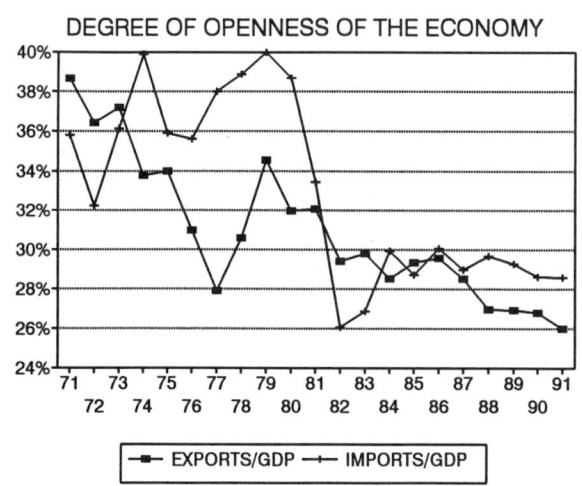

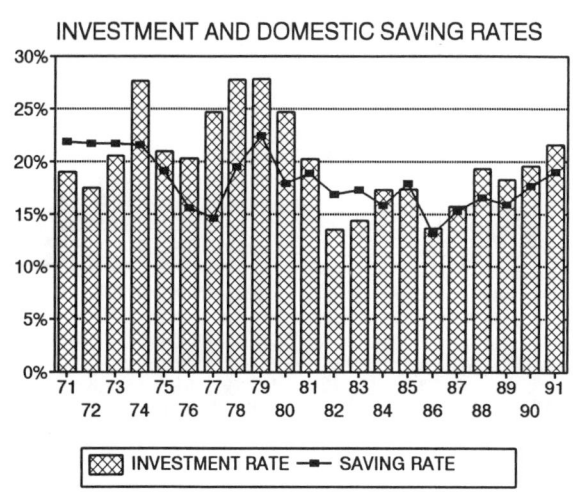

Statistics and Quantitative Analysis/IDB

# JAMAICA
## National Accounts in 1988 US Dollars

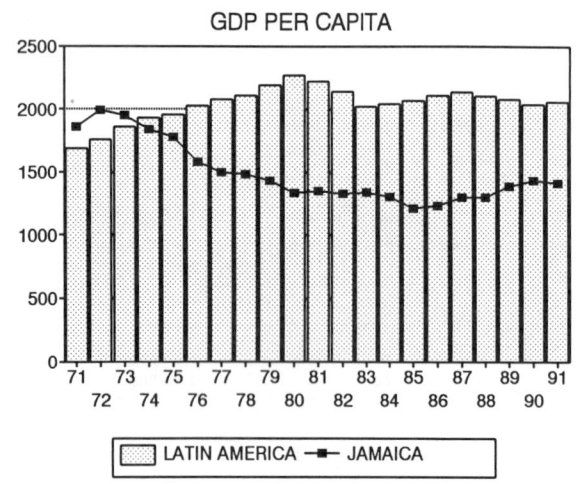

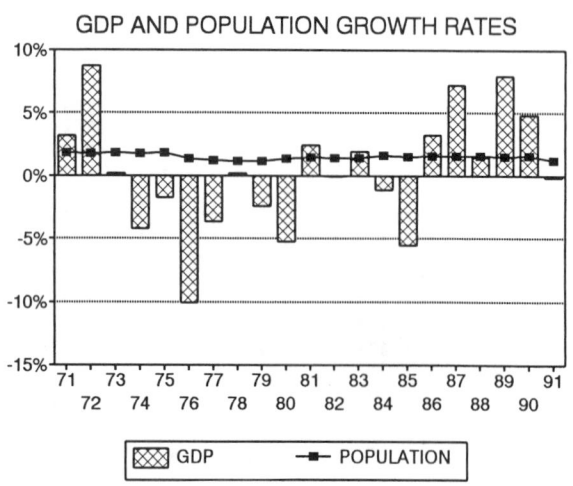

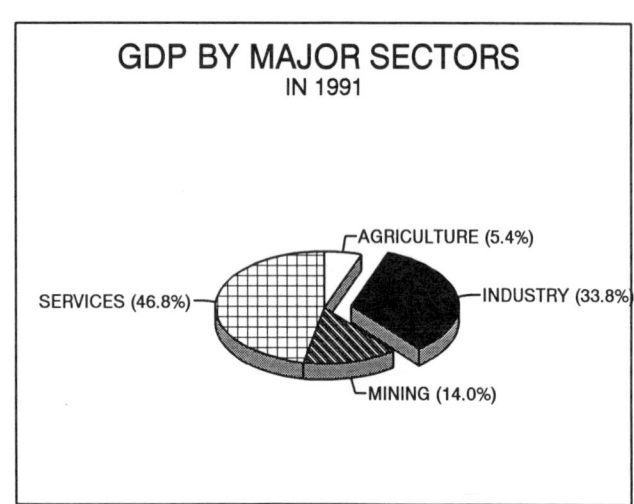

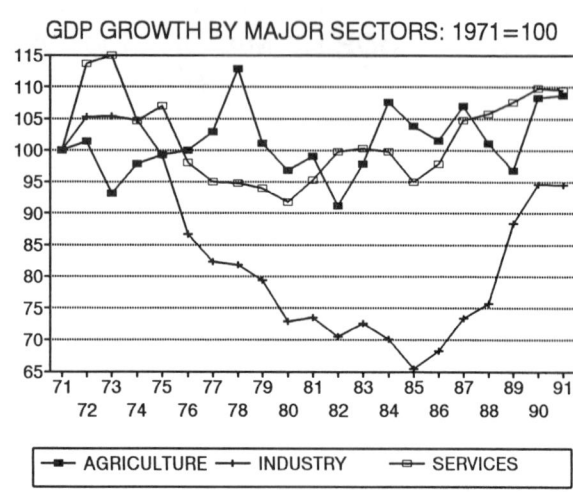

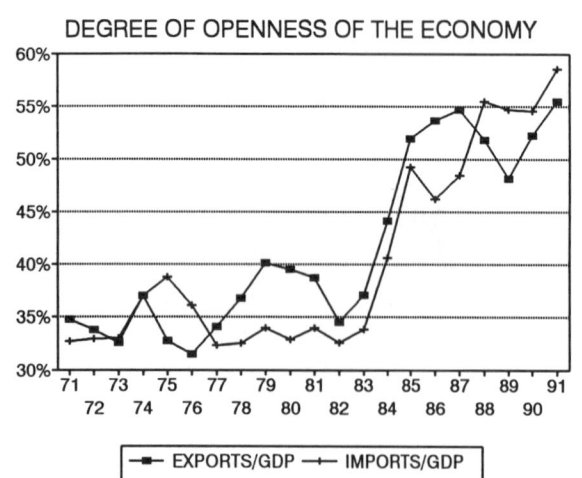

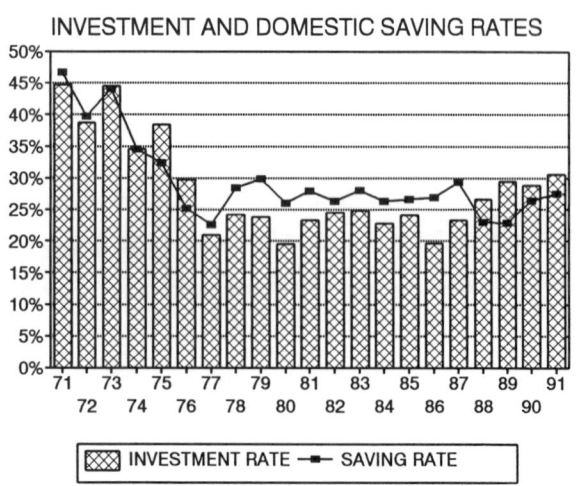

Statistics and Quantitative Analysis/IDB

# MEXICO
## National Accounts in 1988 US Dollars

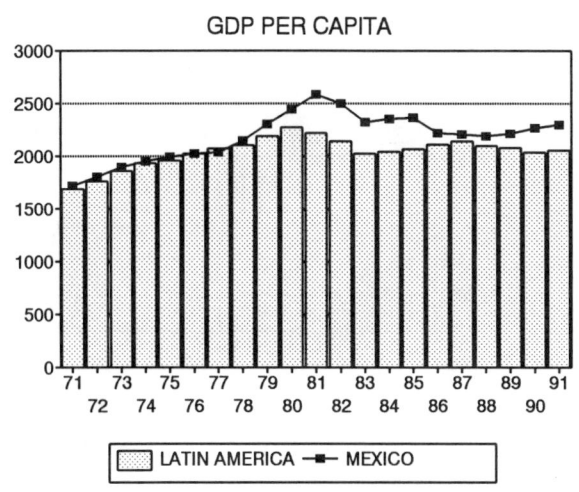

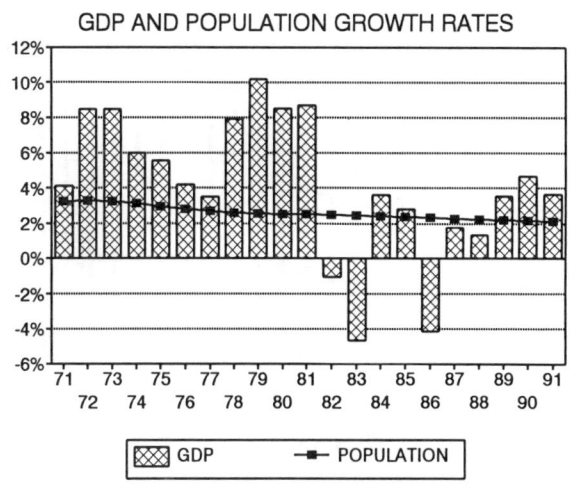

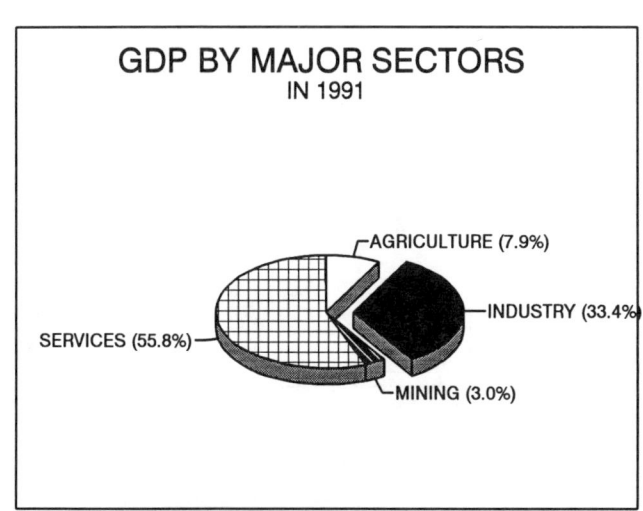

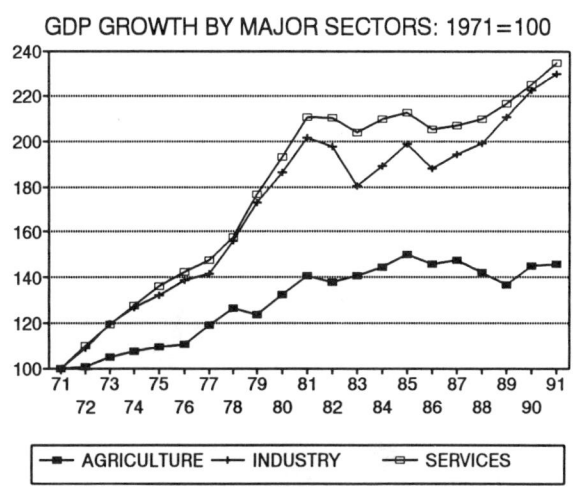

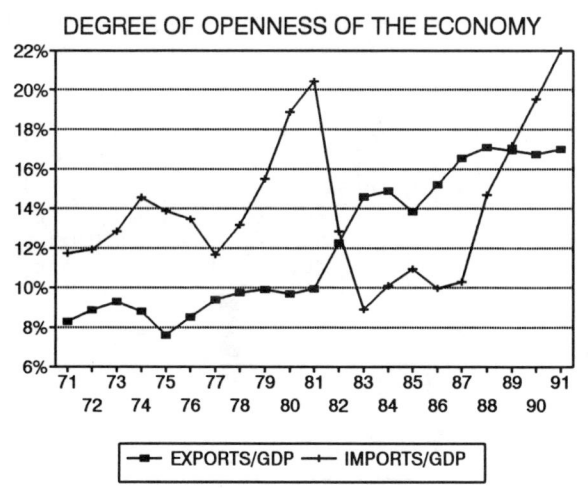

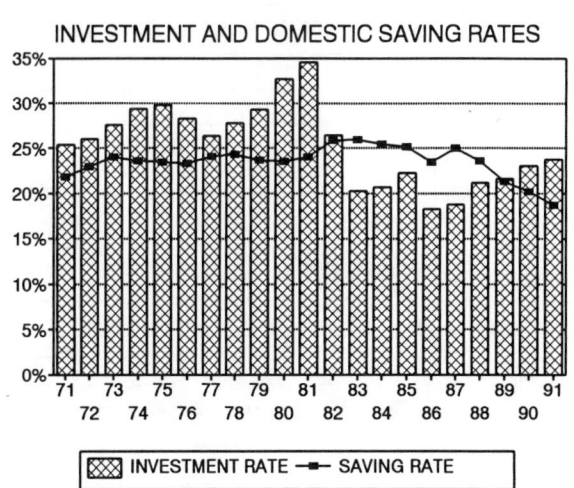

Statistics and Quantitative Analysis/IDB

# NICARAGUA
## National Accounts in 1988 US Dollars

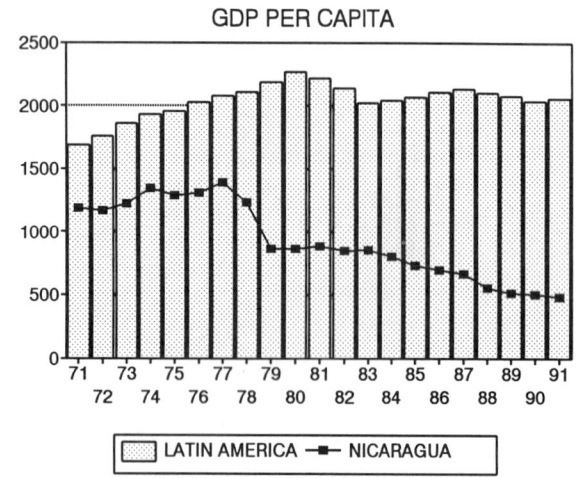

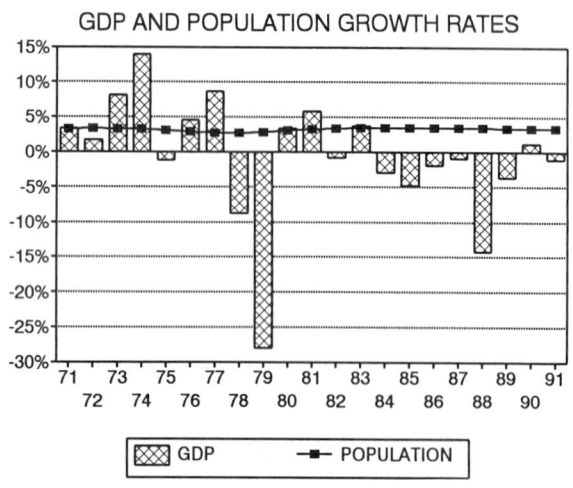

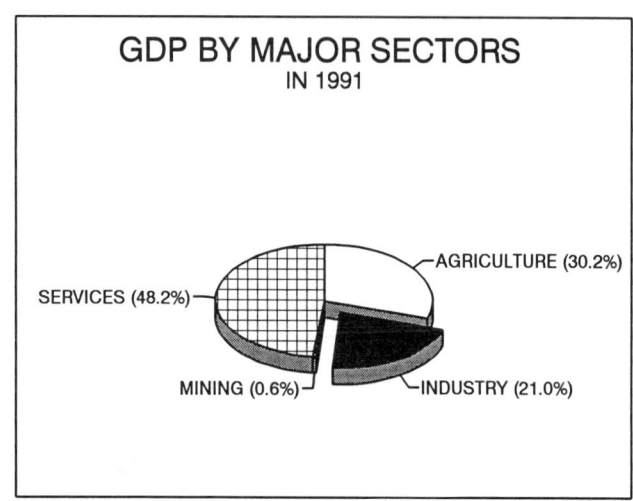

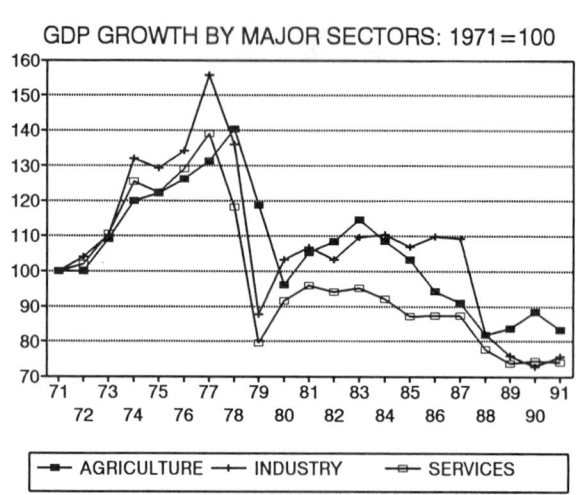

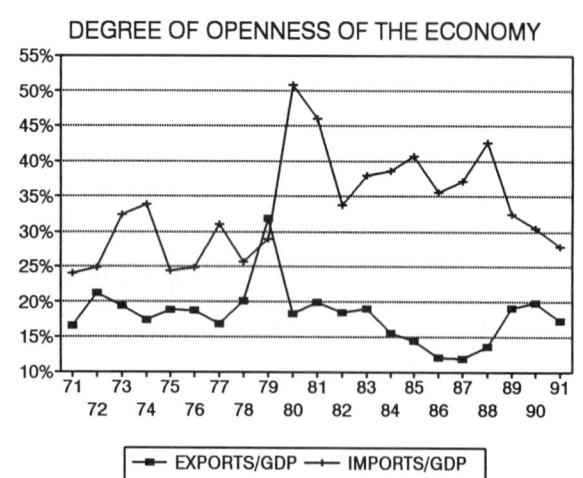

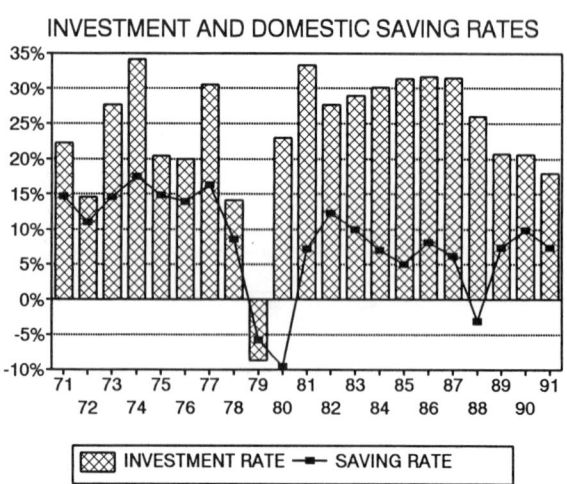

Statistics and Quantitative Analysis/IDB

# PANAMA
## National Accounts in 1988 US Dollars

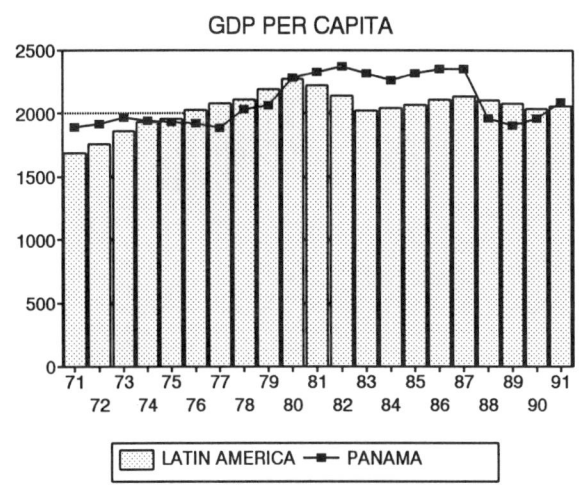

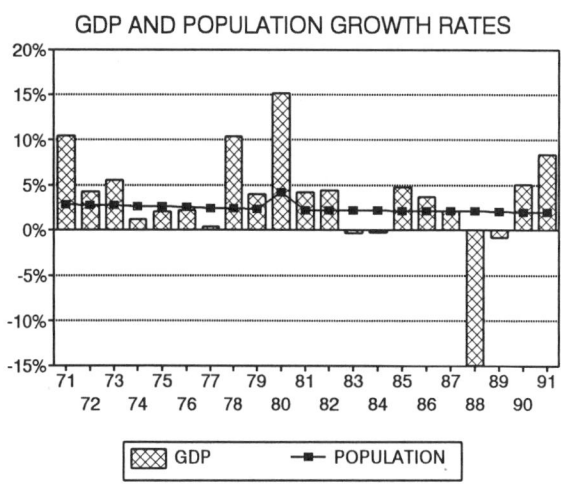

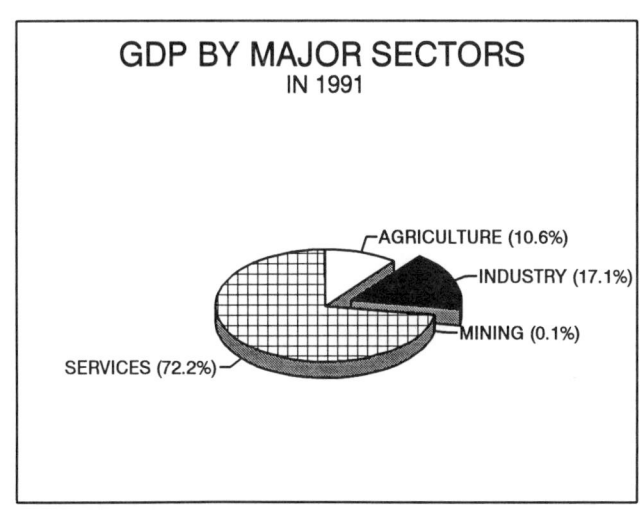

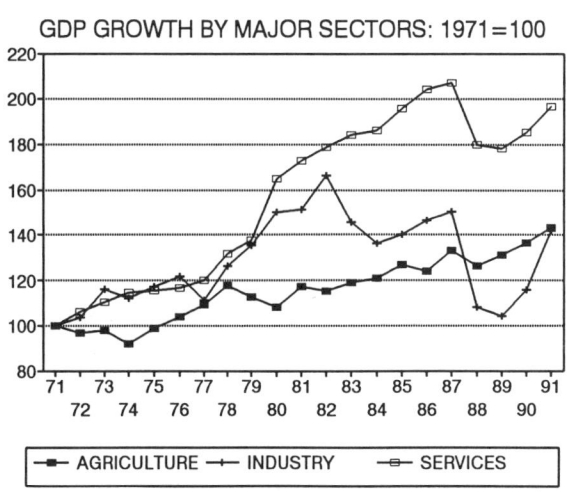

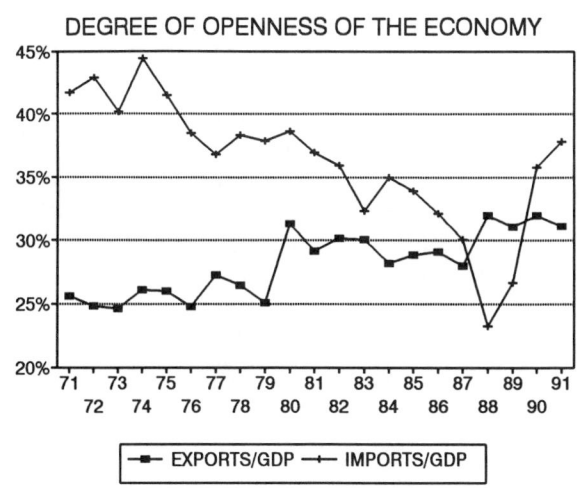

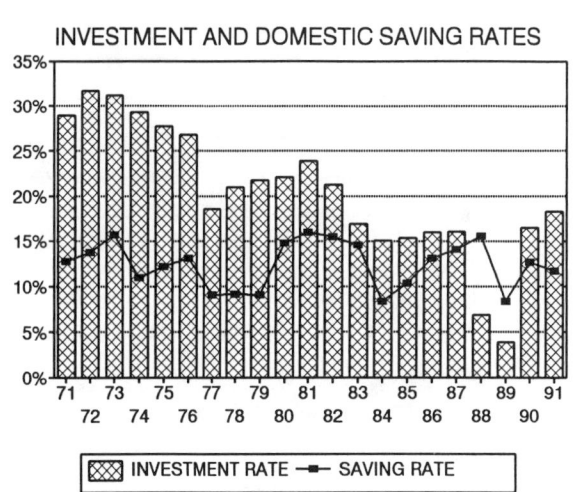

Statistics and Quantitative Analysis/IDB

# PARAGUAY
## National Accounts in 1988 US Dollars

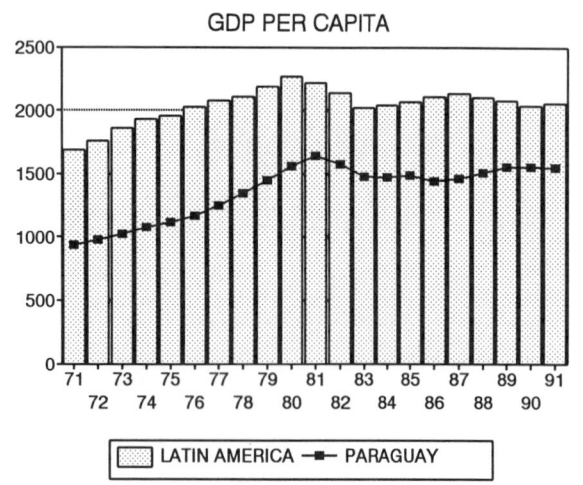

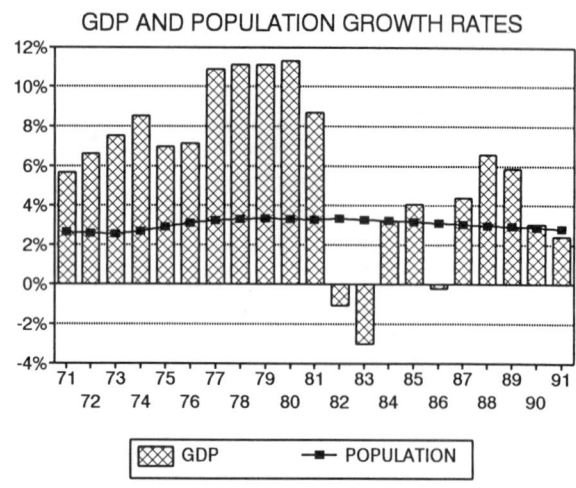

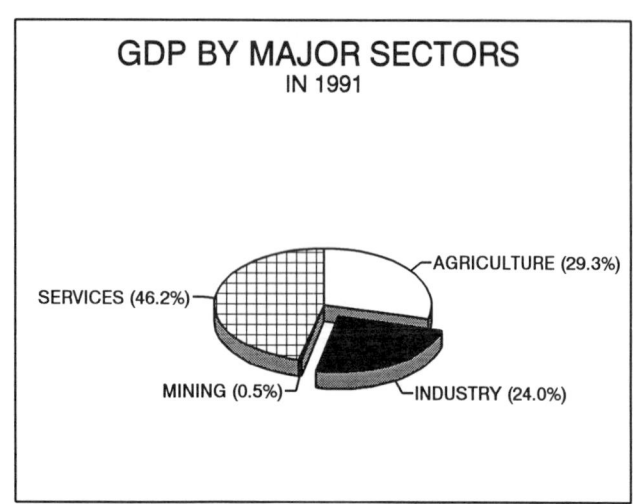

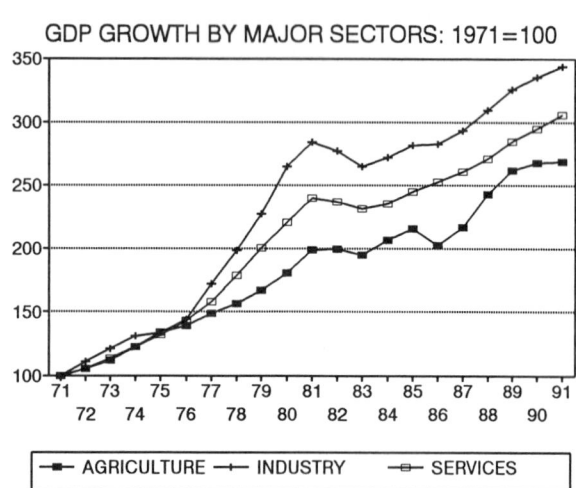

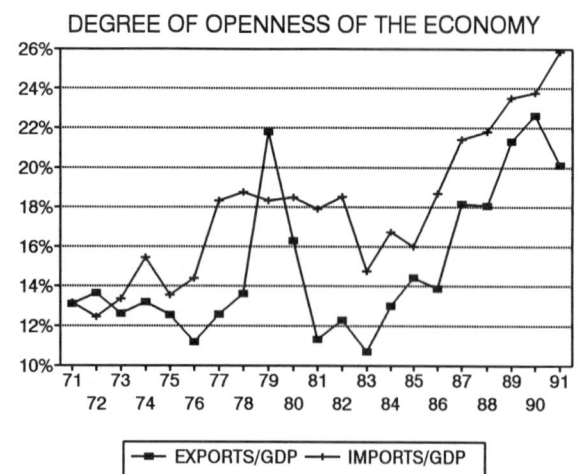

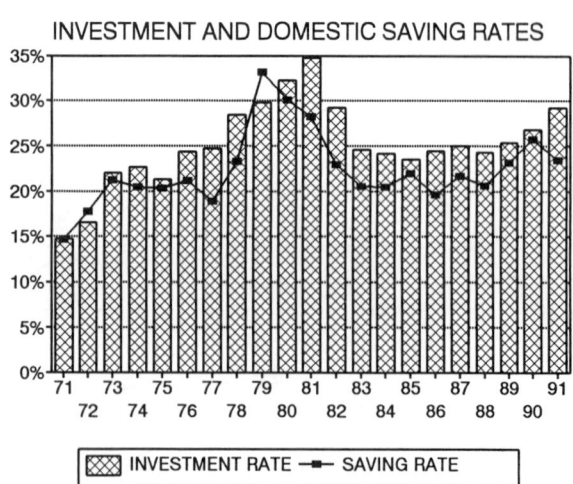

Statistics and Quantitative Analysis/IDB

# PERU
## National Accounts in 1988 US Dollars

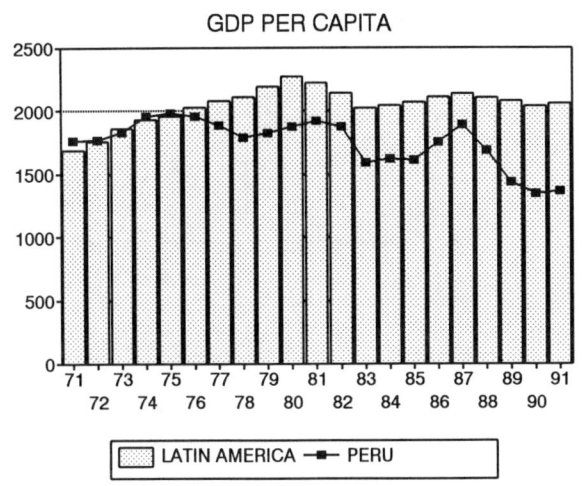

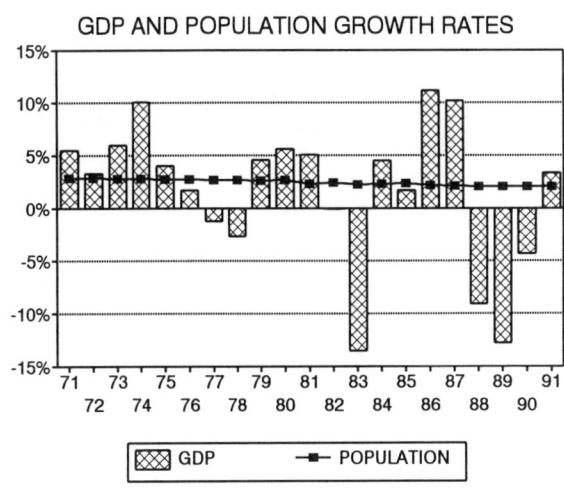

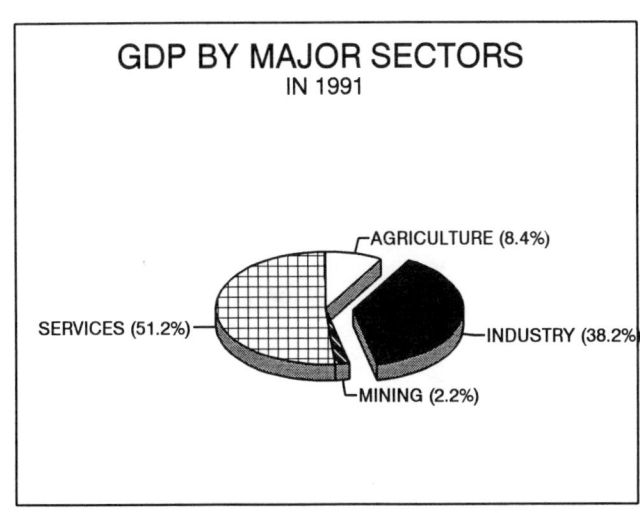

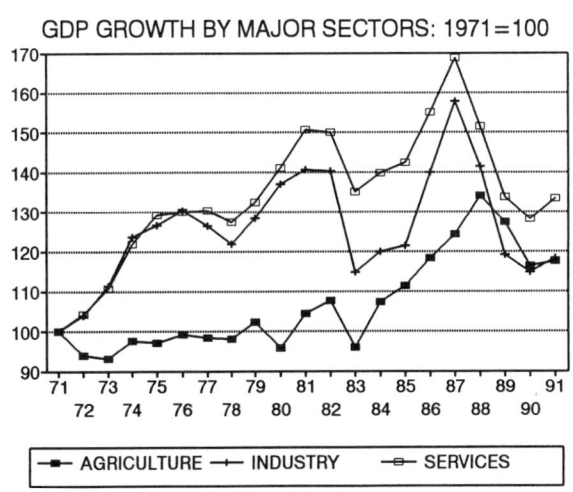

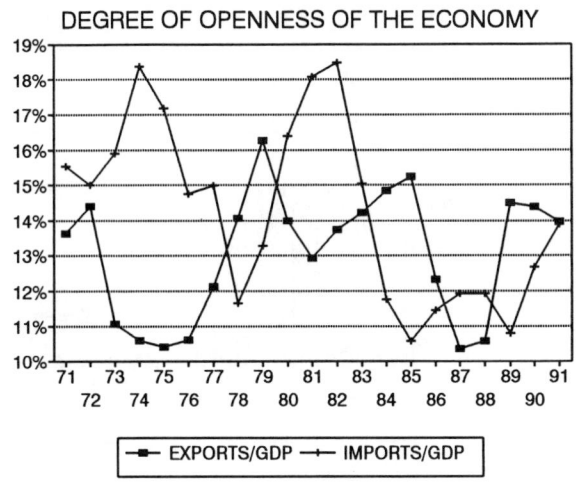

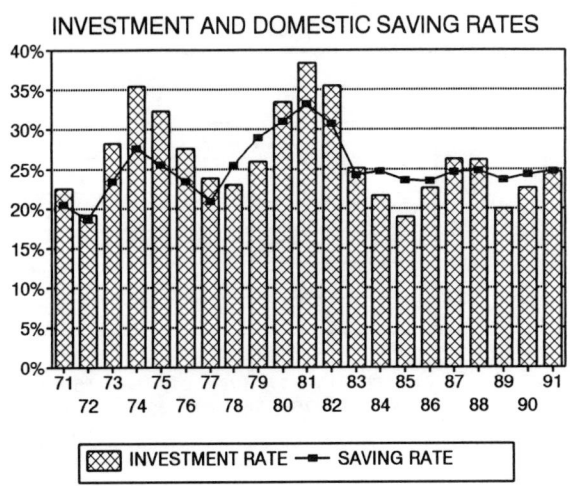

Statistics and Quantitative Analysis/IDB

# SURINAME
## National Accounts in 1988 US Dollars

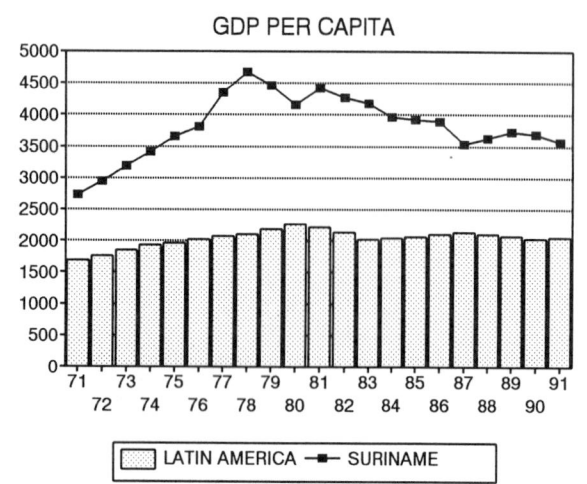

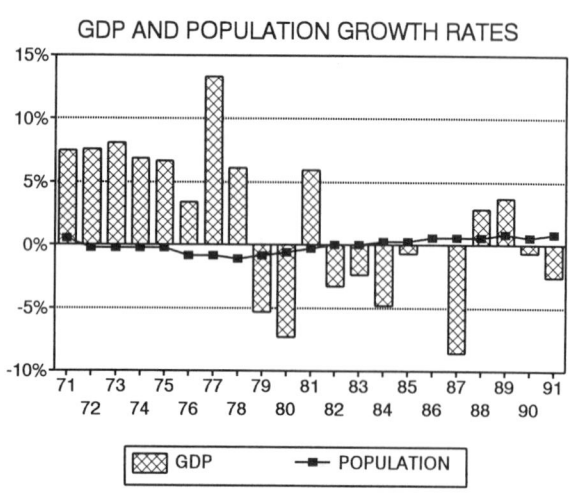

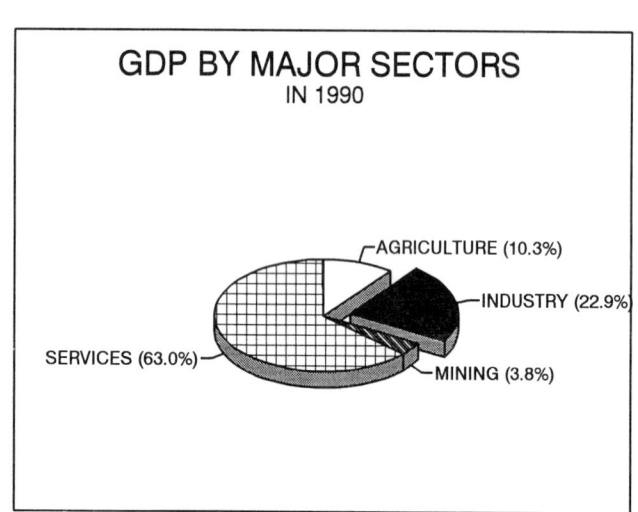

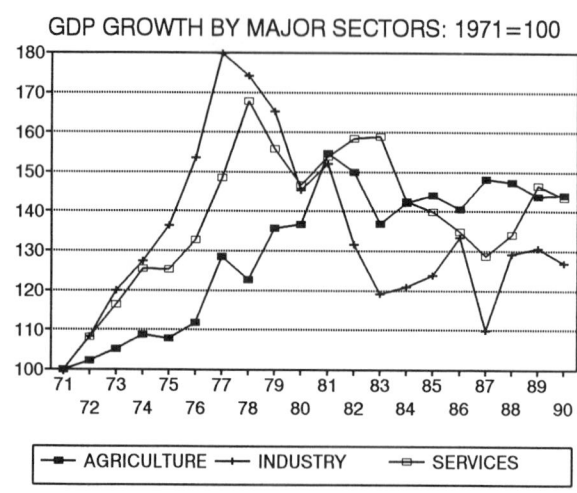

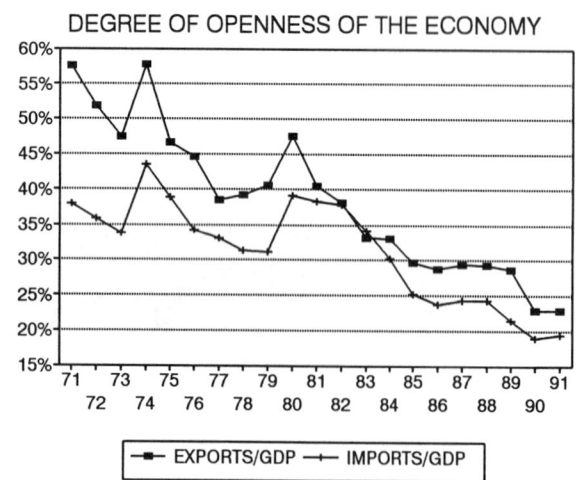

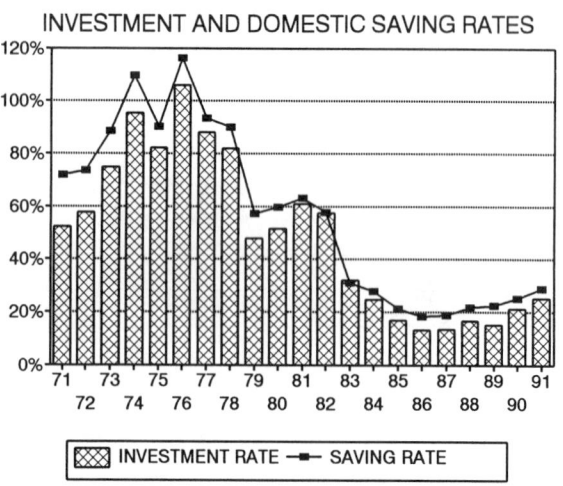

Statistics and Quantitative Analysis/IDB

# TRINIDAD AND TOBAGO
## National Accounts in 1988 US Dollars

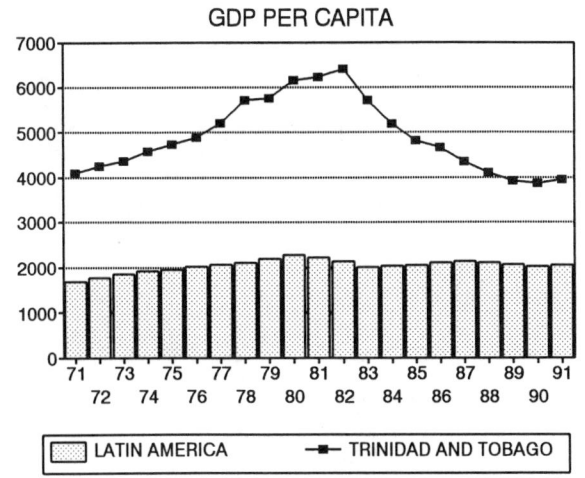

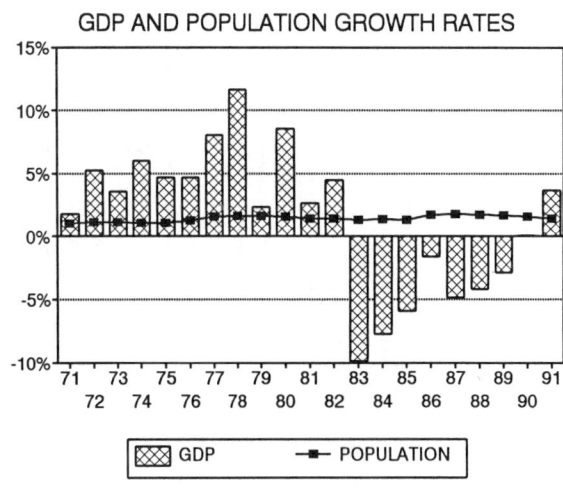

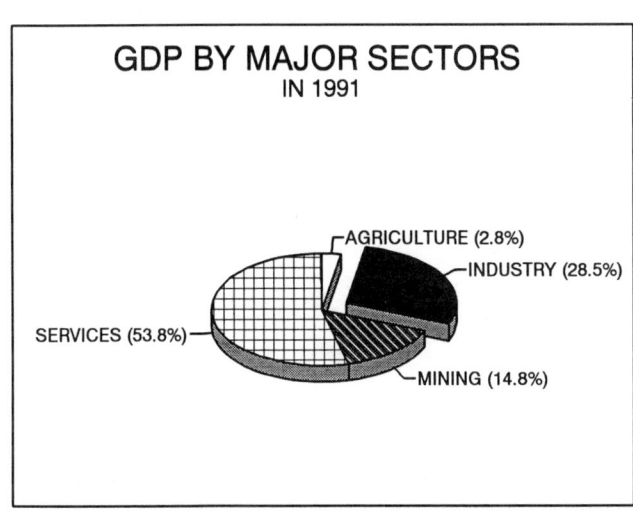

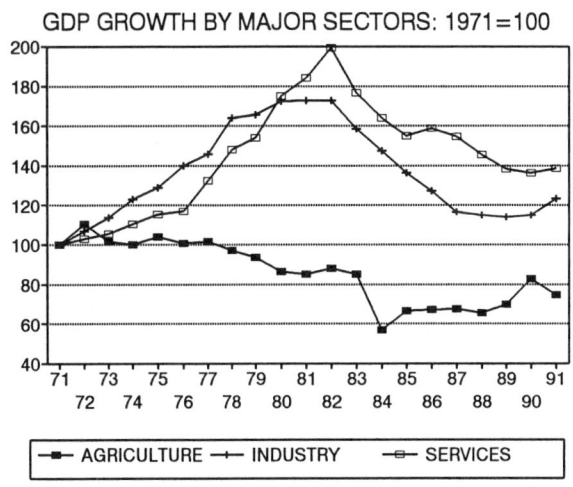

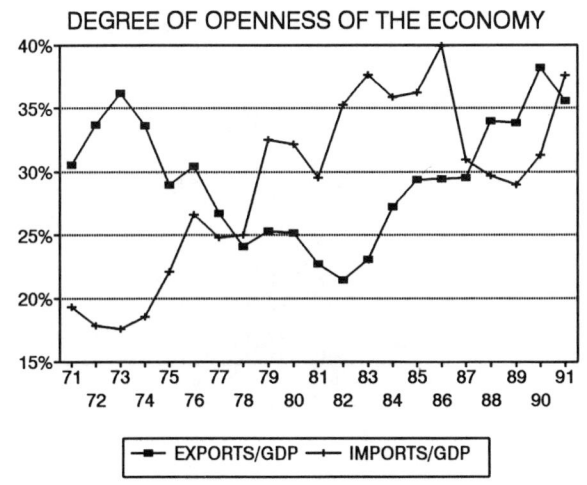

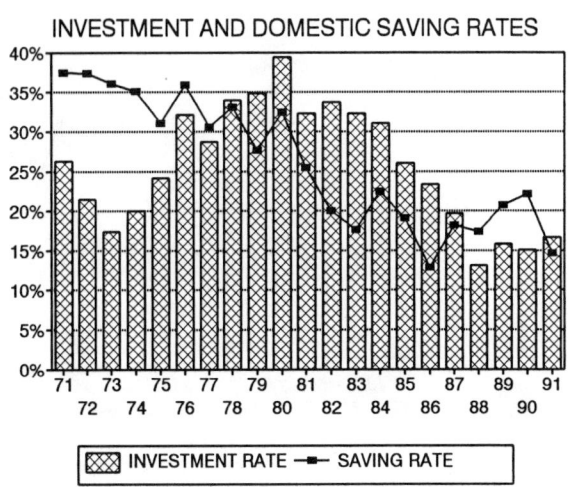

Statistics and Quantitative Analysis/IDB

# URUGUAY
## National Accounts in 1988 US Dollars

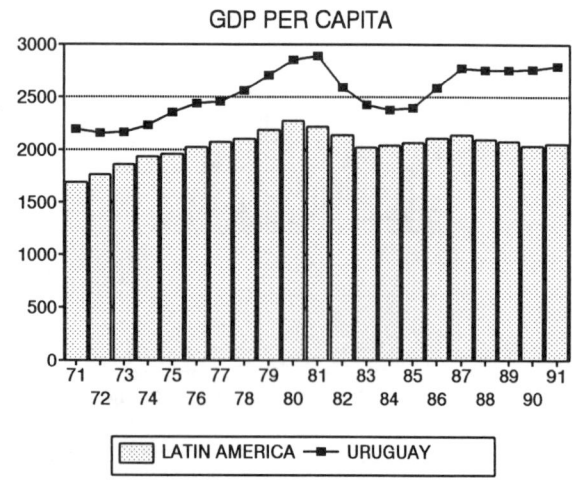
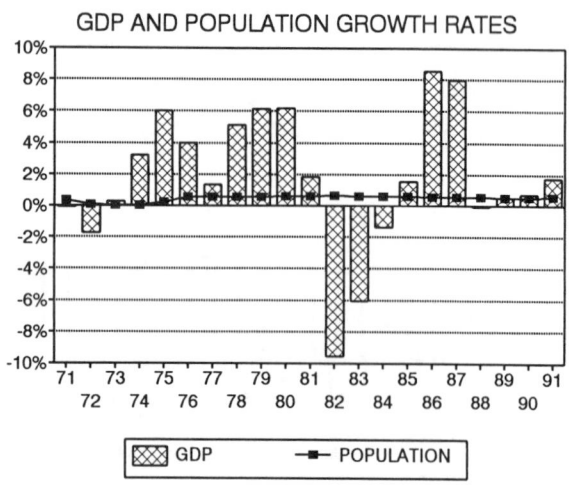
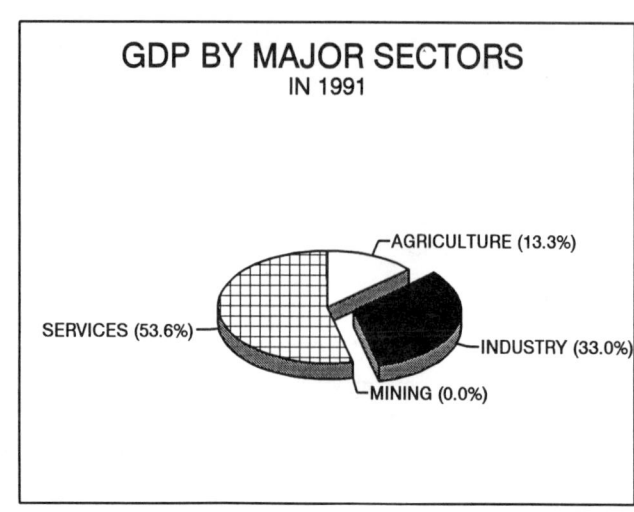
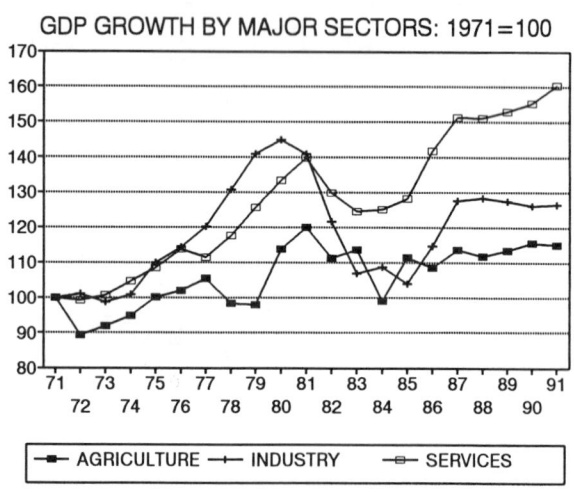
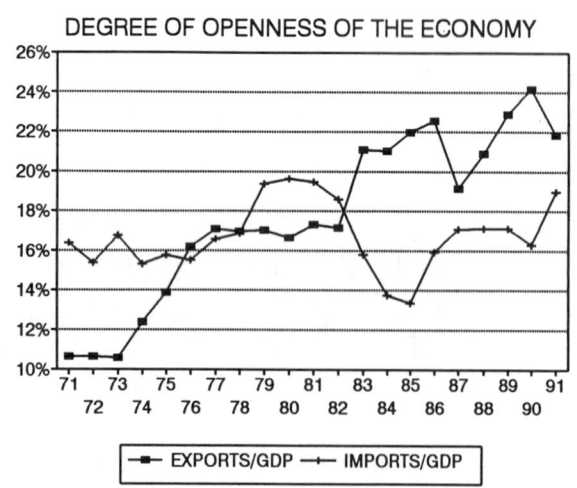
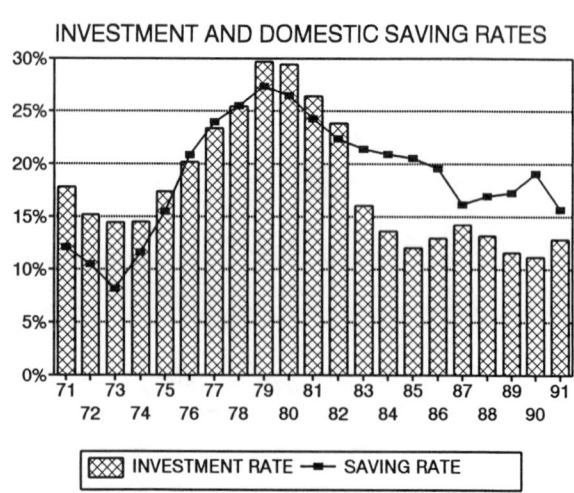

Statistics and Quantitative Analysis/IDB

# VENEZUELA
## National Accounts in 1988 US Dollars

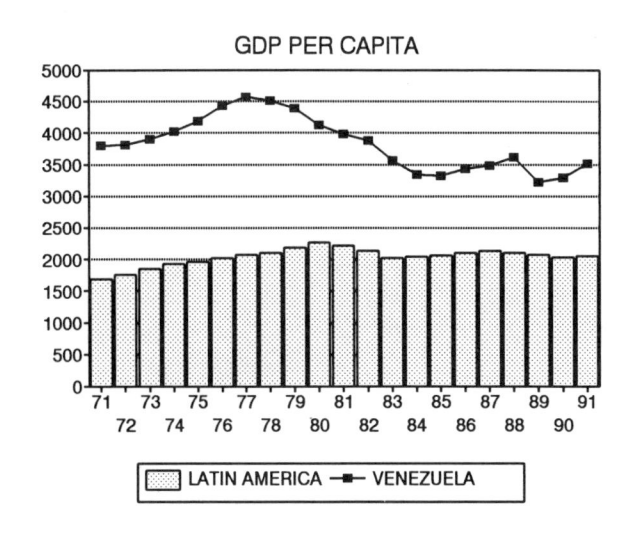

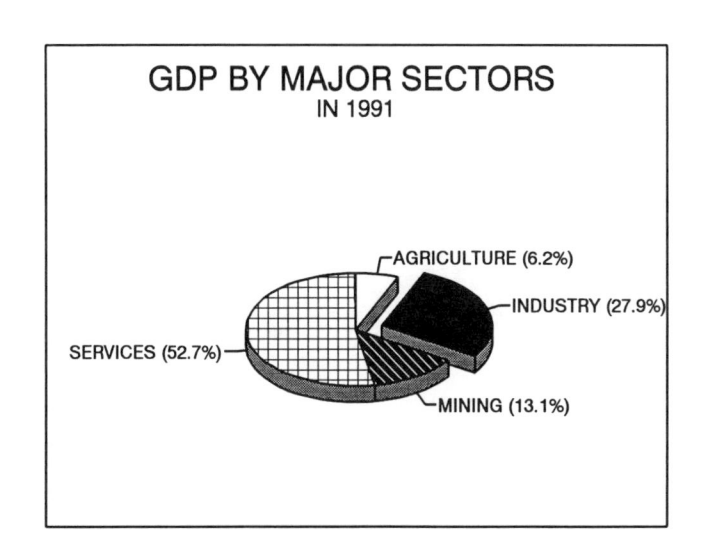

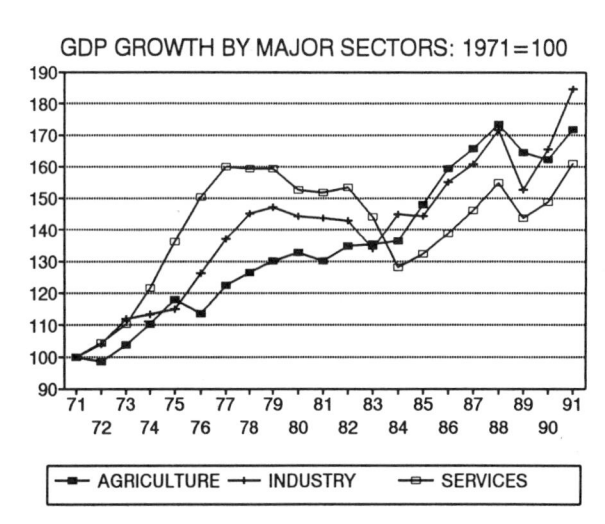

### GDP PER CAPITA
(LATIN AMERICA, VENEZUELA)

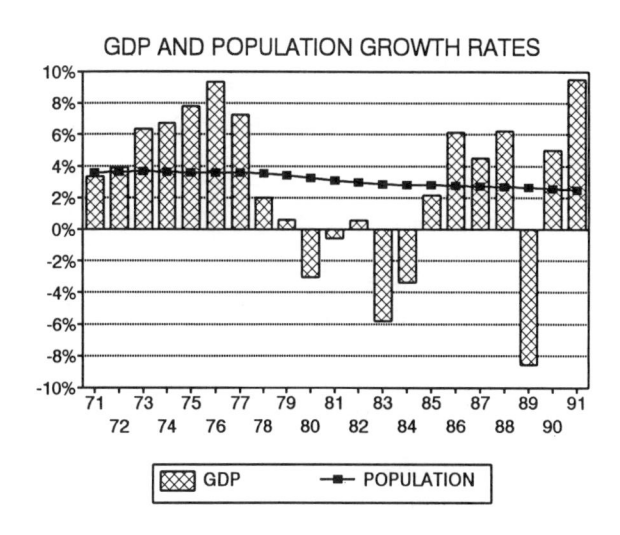

### GDP AND POPULATION GROWTH RATES
(GDP, POPULATION)

### GDP BY MAJOR SECTORS IN 1991
- AGRICULTURE (6.2%)
- INDUSTRY (27.9%)
- MINING (13.1%)
- SERVICES (52.7%)

### GDP GROWTH BY MAJOR SECTORS: 1971=100
(AGRICULTURE, INDUSTRY, SERVICES)

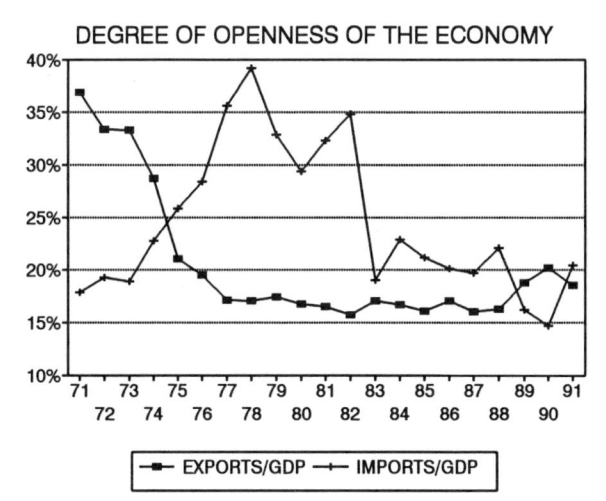

### DEGREE OF OPENNESS OF THE ECONOMY
(EXPORTS/GDP, IMPORTS/GDP)

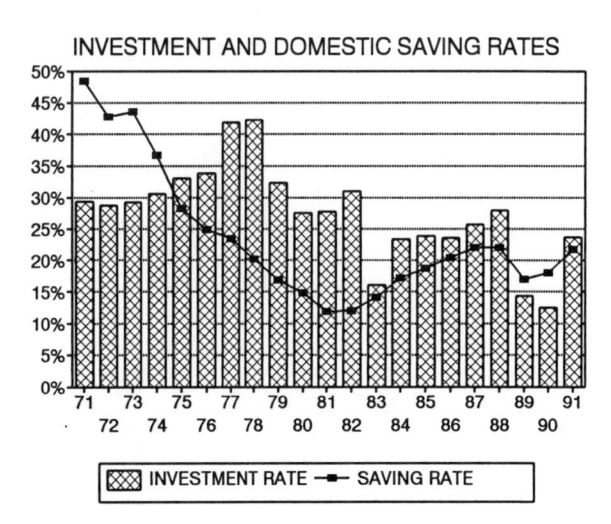

### INVESTMENT AND DOMESTIC SAVING RATES
(INVESTMENT RATE, SAVING RATE)

Statistics and Quantitative Analysis/IDB

## Country Notes on National Accounts

### Bahamas
Since GDP is the only national accounts series to be estimated in constant prices by the local authorities, only the first two graphs are available for the country.

### Barbados
Investment is Gross Fixed Capital Formation only.

### Brazil
Beginning in 1985, Investment excludes the Change in Stocks.

### Costa Rica
Mining and Quarrying is included in Manufacturing.

### Guyana
Electricity, Gas and Water is included in Services.

### Suriname
The Structure of GDP by Major Sectors is shown for 1990 as data are not available for 1991.

### Uruguay
Mining and Quarrying is included in Manufacturing.

# BALANCE OF PAYMENTS

## BALANCE OF PAYMENTS

Data are from the International Monetary Fund Balance of Payments tapes and IDB estimates based on official statistics for terminal years.

### CURRENT ACCOUNT BALANCE

GDP in current dollars is calculated by using IDB estimated conversion factors. Expressed as a percentage of GDP, a Current Account deficit indicates the degree of external financing requirements. Unrequited Transfers from abroad (mainly workers remittances and official grants) are not considered as external financing since they are part of the Current Account Balance. As an example, in the case of Haiti where grants are important, the degree of external financing appears low in comparison with other countries.

### CURRENT ACCOUNT COMPONENTS AS A PERCENT OF GDP

The graph shows a breakdown of the Current Account Balance between the balance of Goods, Non-Financial Services (NFS) and Transfers on the one hand and Net Investment Income on the other. Net Investment Income includes interest (on an accrued basis), dividends on shares and profits of enterprises. It appears almost always negative, reflecting the burden of external debt. The progressive drop in international interest rates since 1982 together with the reduction of external debt as a percent of GDP in the last few years, explains the decrease in interest payments in many countries.

### EXPORTS AND IMPORTS OF GOODS FOB

The graph shows the continuing improvement in exports of goods for most countries during the last two decades and, at the same time, the drastic reduction in imports in the aftermath of the crisis of 1981-1982.

### PRIVATE AND PUBLIC CAPITAL FLOWS

Private capital flows are inflows and outflows from the Private Sector of the country. The Private Sector is that part of the economy most likely to be affected by market forces. It can be described as the Enterprise Sector (and includes Government-owned Enterprises). Public capital flows are inflows and outflows from the Public Sector, which include the Government and the Central Monetary Institutions.

Bonds issued by the Government to cancel external obligations or within debt-bond swap operations are not considered as Portfolio Investment (contrary to the presentation in the balance of payments) but rather are included in the long-term operations of the Government (e.g., Argentina in 1982-90, Mexico in 1988-90 and Venezuela in 1990). In fact, these operations are only changing the amount and the nature of the liabilities of the Government sector.

### NET DIRECT INVESTMENT

Net Direct Investment (including re-invested earnings) is expressed as a percent of GDP. It may be negative when some national companies invest their capital abroad or when foreign companies repatriate their capital (e.g., because of nationalization). Participation of foreign capital in national companies is considered as Direct Investment when their purchase of shares give them a controlling voice in the enterprise.

### CHANGE IN INTERNATIONAL RESERVES

Change in International Reserves is calculated in months of Imports of Goods. In some cases, the recent accumulation of International Reserves has reached the equivalent of several months of Imports.

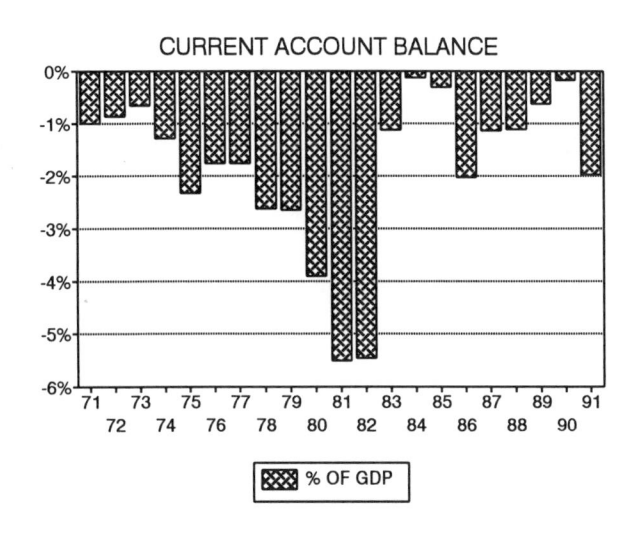

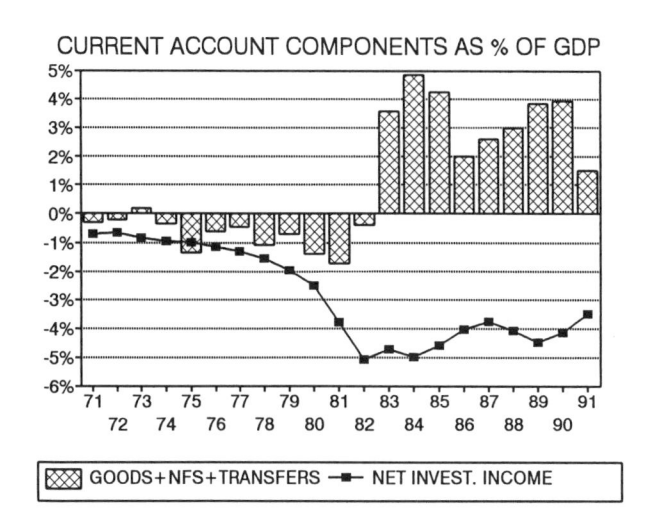

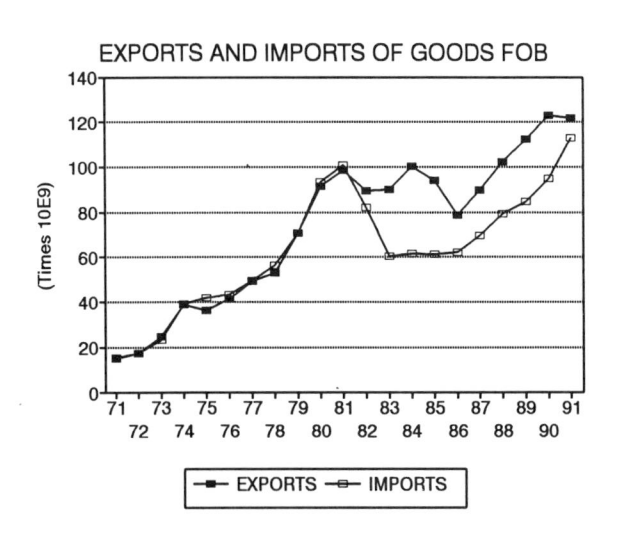

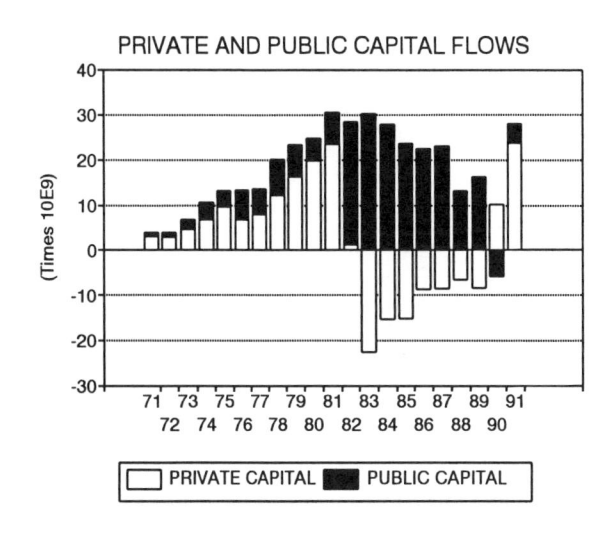

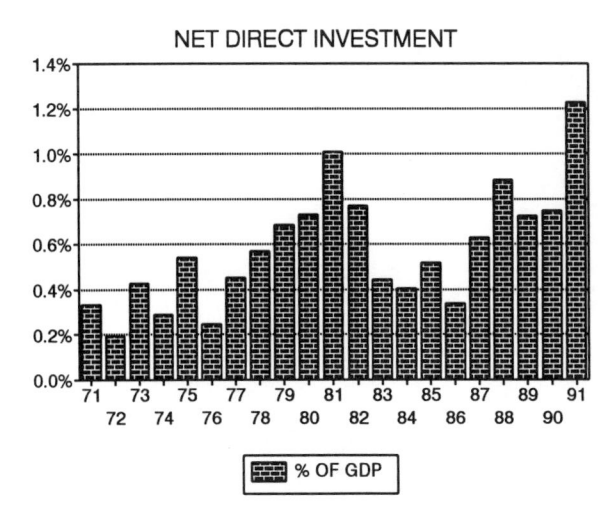

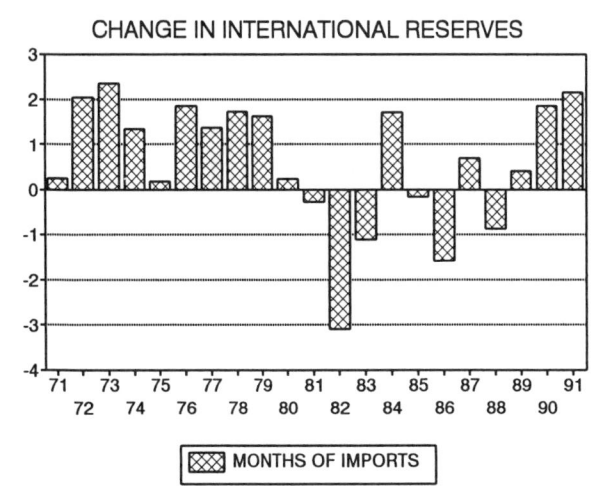

# LATIN AMERICA
## Balance of Payments in US Dollars

Statistics and Quantitative Analysis/IDB

# ARGENTINA
## Balance of Payments in US Dollars

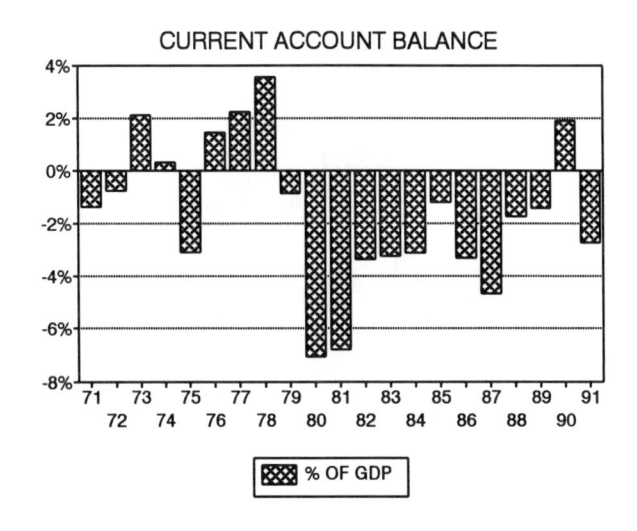

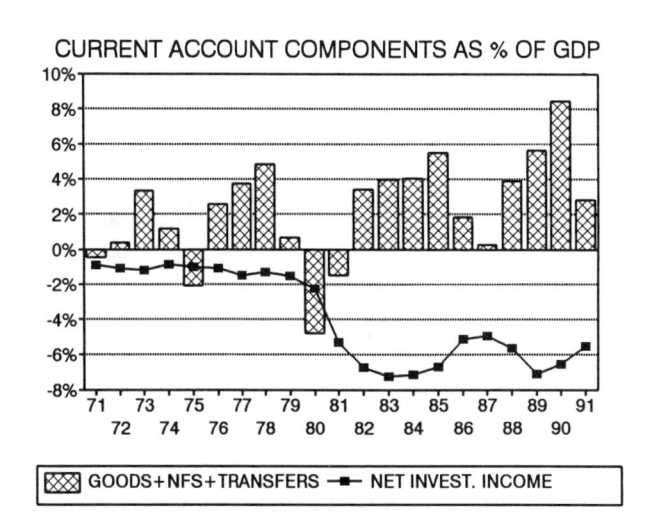

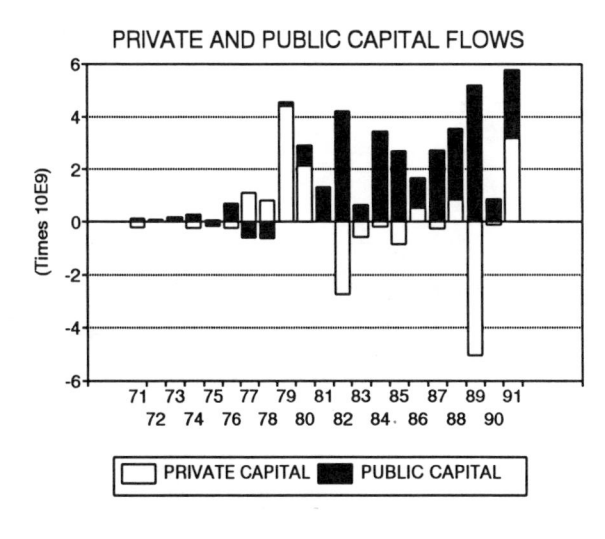

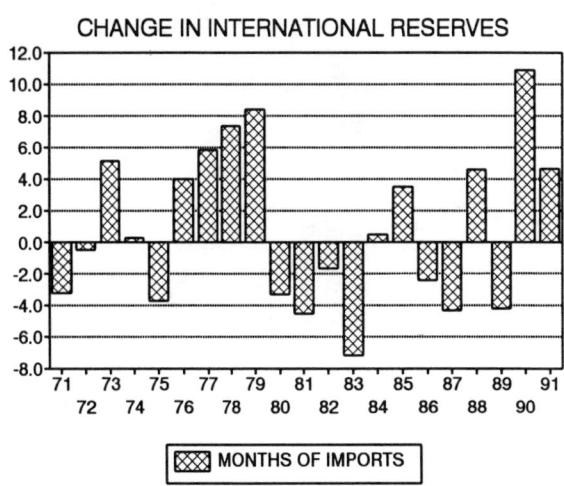

### CURRENT ACCOUNT BALANCE
% OF GDP

### CURRENT ACCOUNT COMPONENTS AS % OF GDP
GOODS+NFS+TRANSFERS — NET INVEST. INCOME

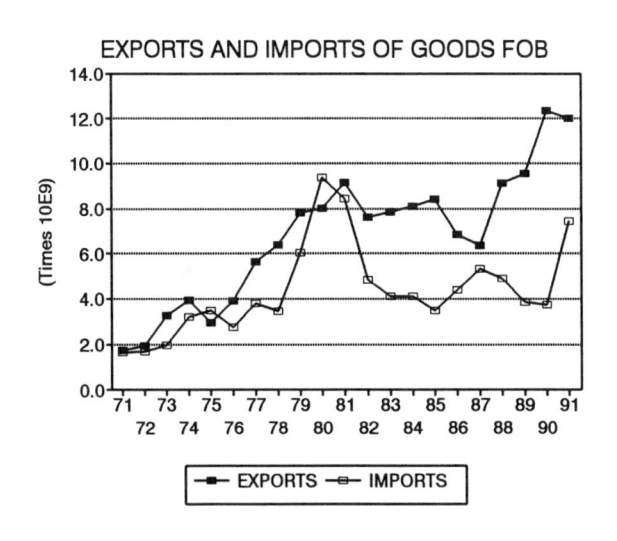

### EXPORTS AND IMPORTS OF GOODS FOB
(Times 10E9)
EXPORTS — IMPORTS

### PRIVATE AND PUBLIC CAPITAL FLOWS
(Times 10E9)
PRIVATE CAPITAL — PUBLIC CAPITAL

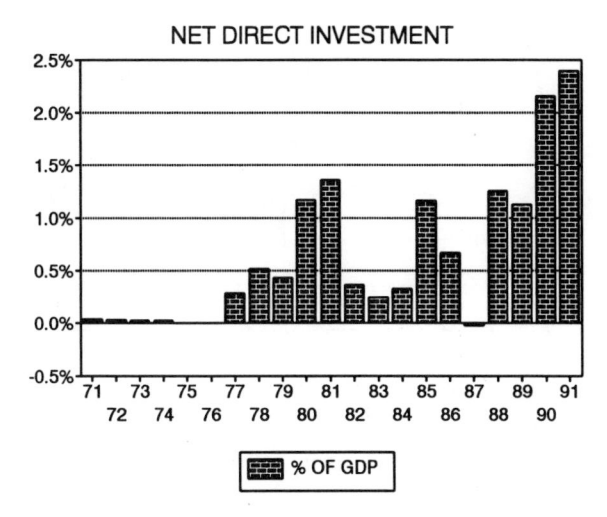

### NET DIRECT INVESTMENT
% OF GDP

### CHANGE IN INTERNATIONAL RESERVES
MONTHS OF IMPORTS

Statistics and Quantitative Analysis/IDB

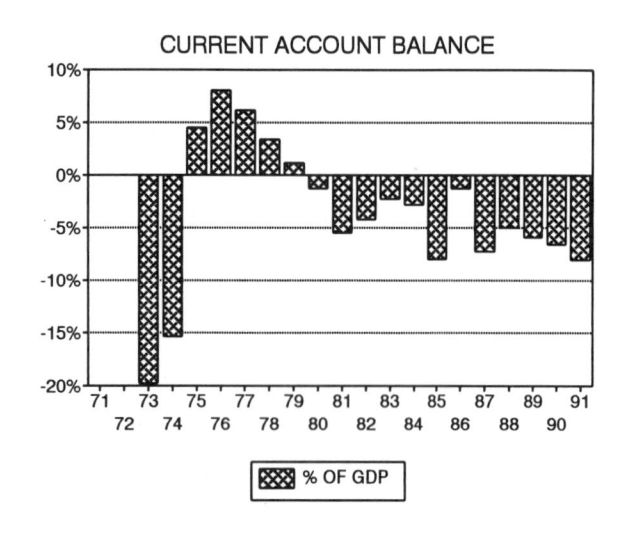

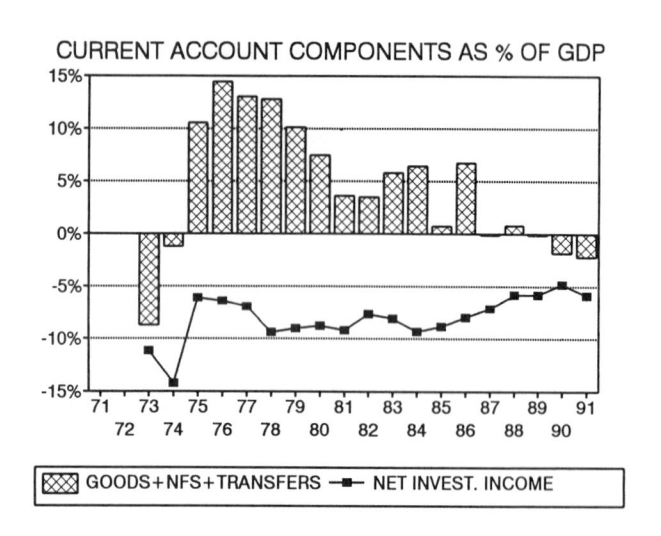

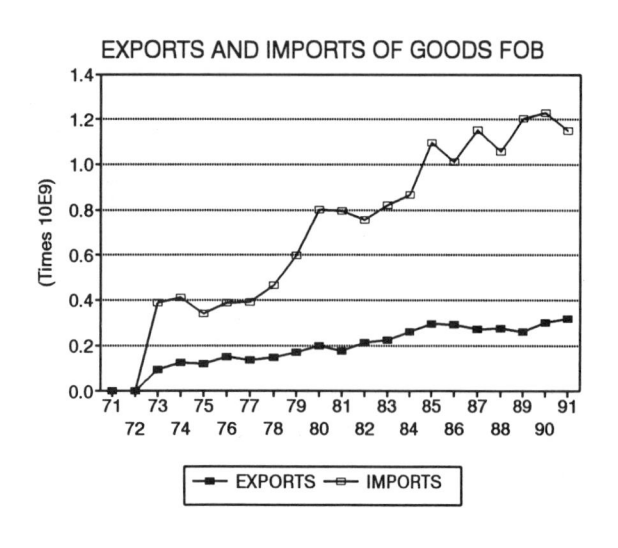

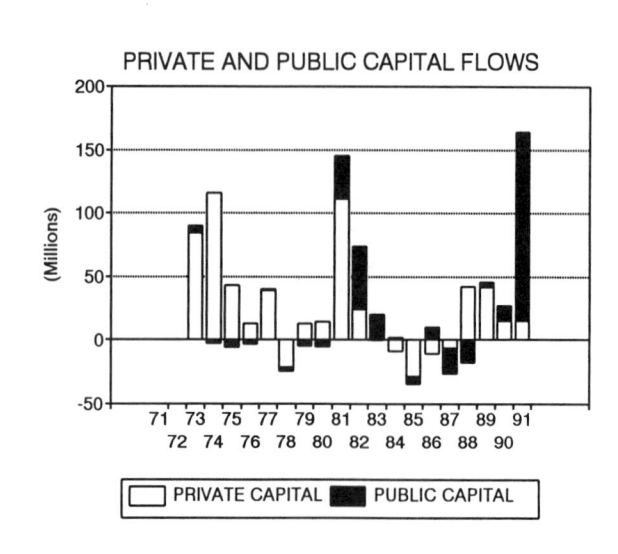

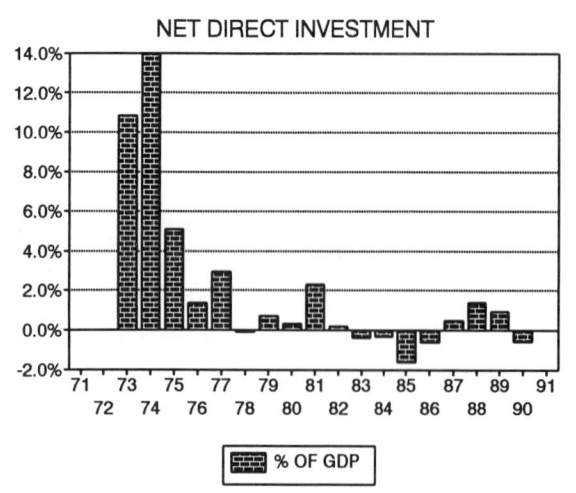

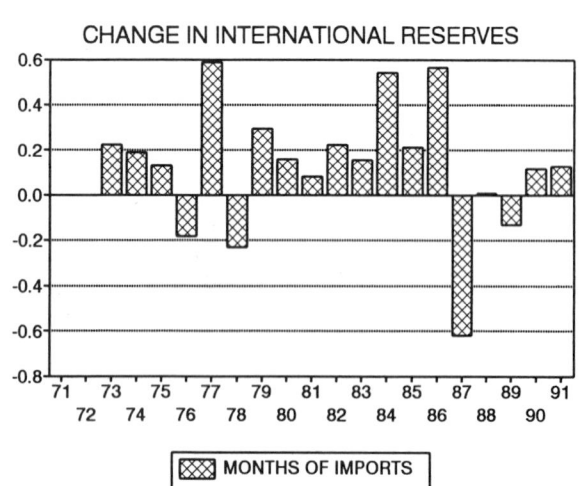

# BAHAMAS
## Balance of Payments in US Dollars

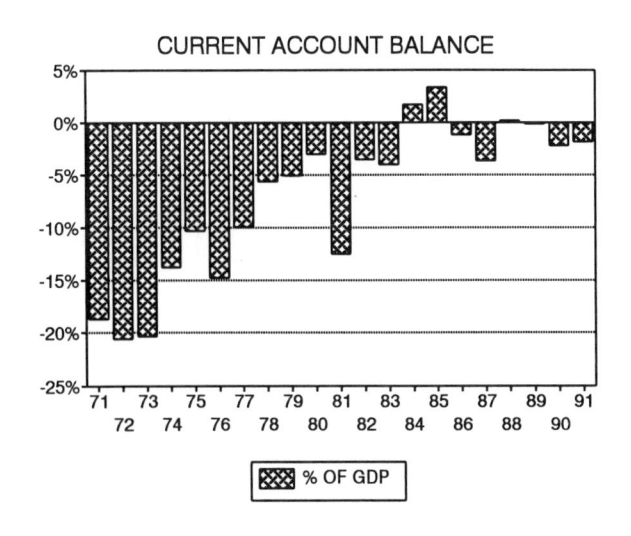

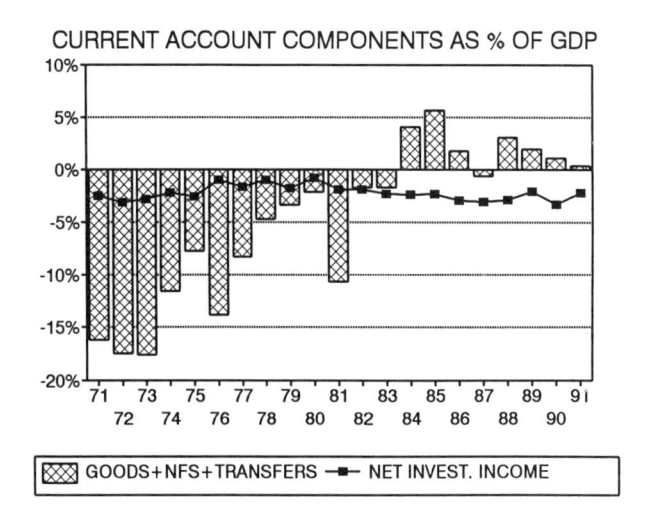

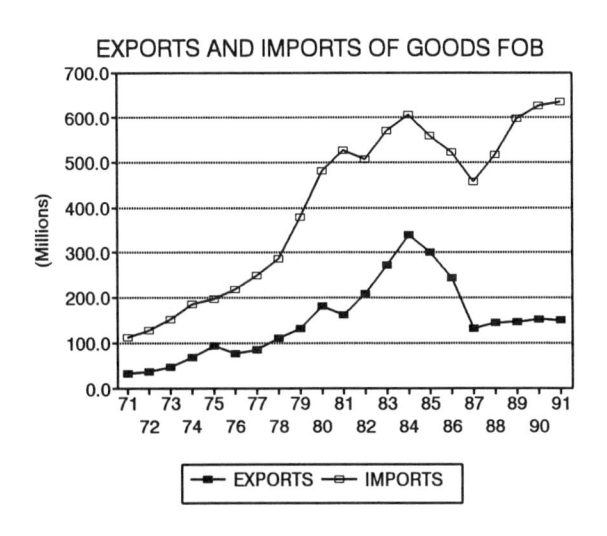

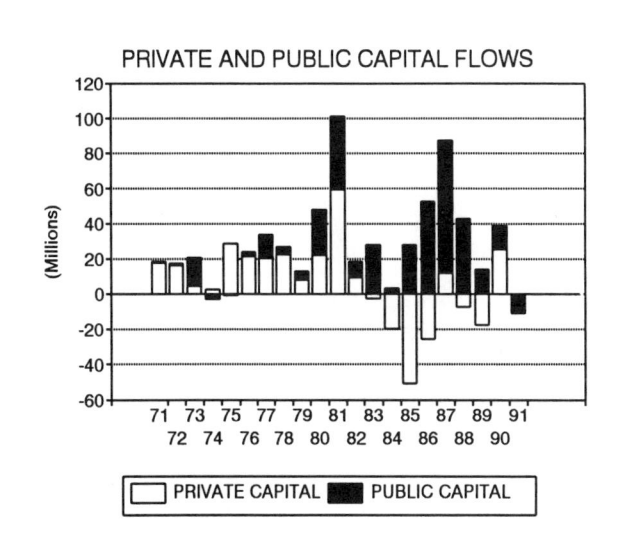

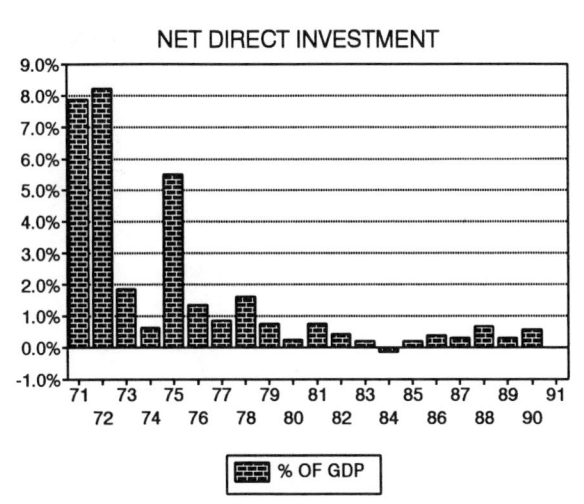

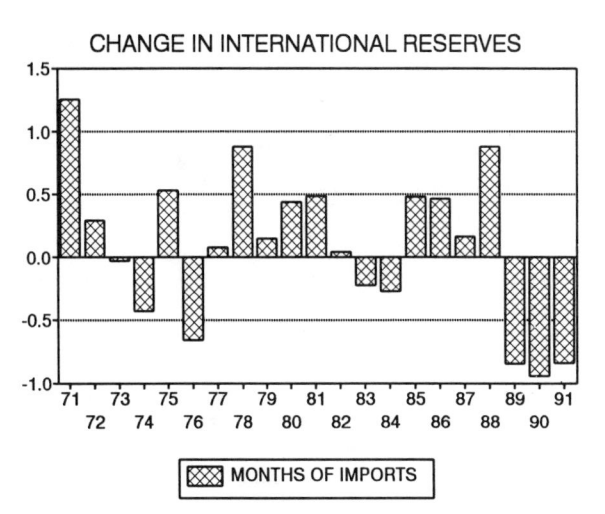

# BARBADOS
## Balance of Payments in US Dollars

Statistics and Quantitative Analysis/IDB

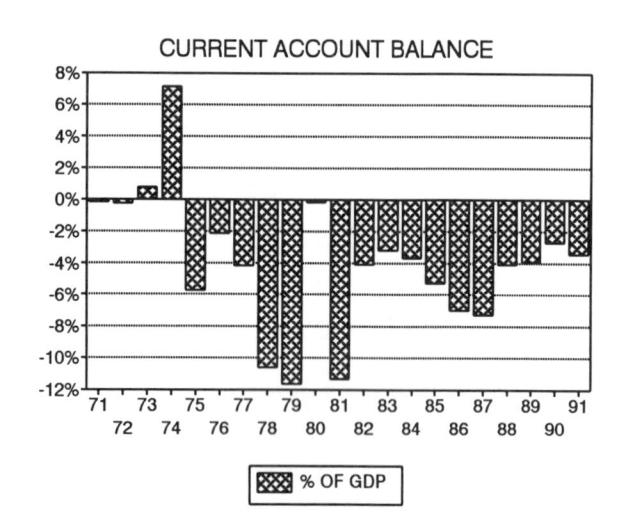

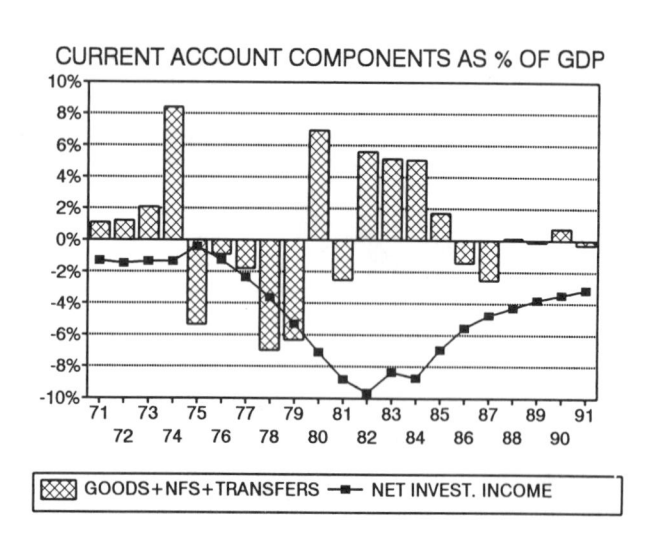

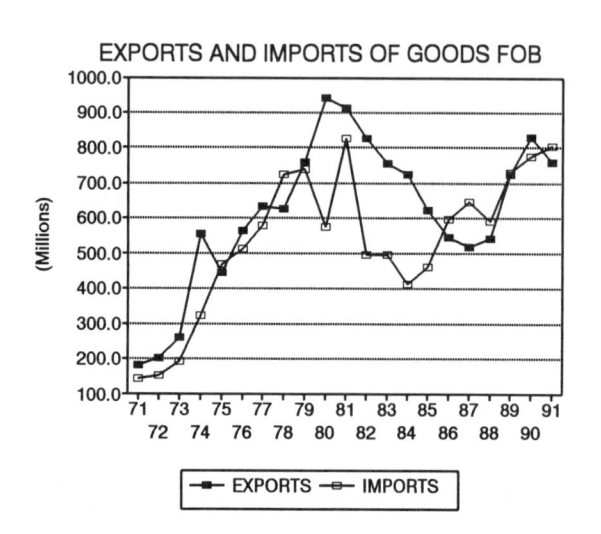

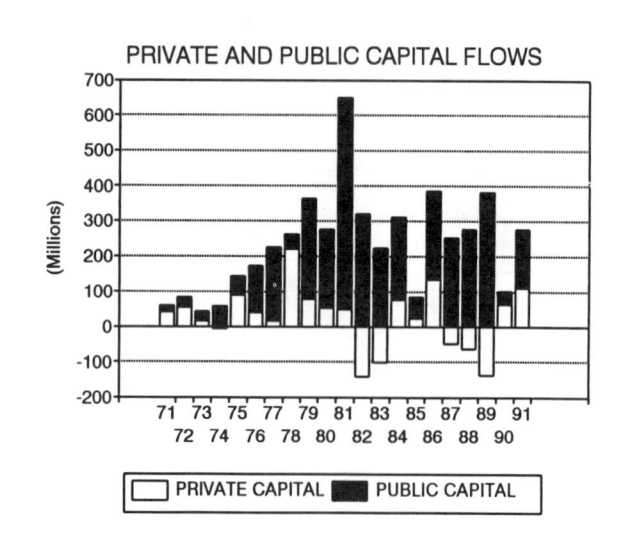

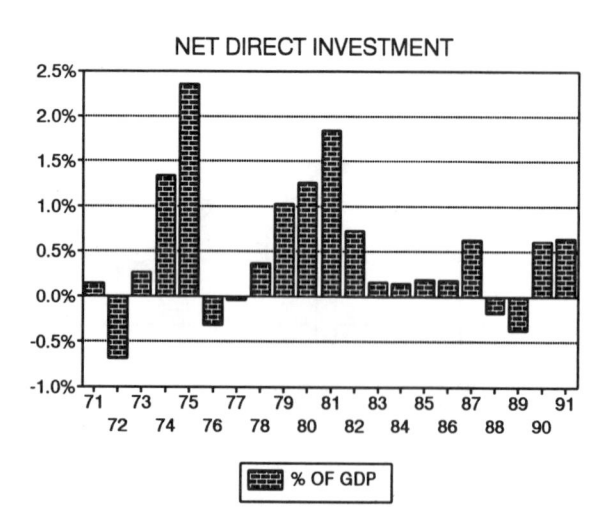

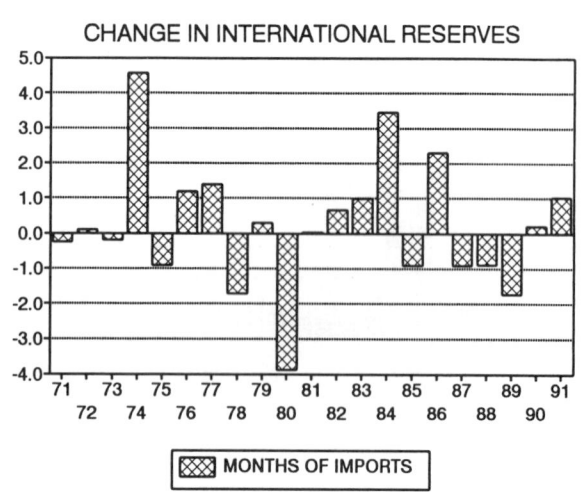

# BOLIVIA
## Balance of Payments in US Dollars

# BRAZIL
## Balance of Payments in US Dollars

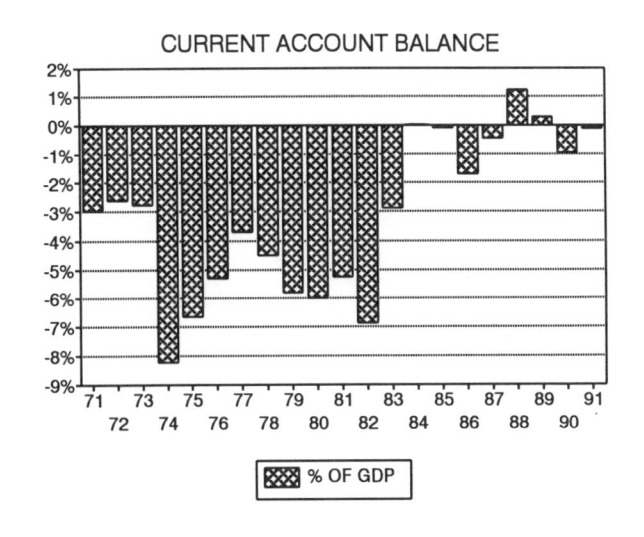

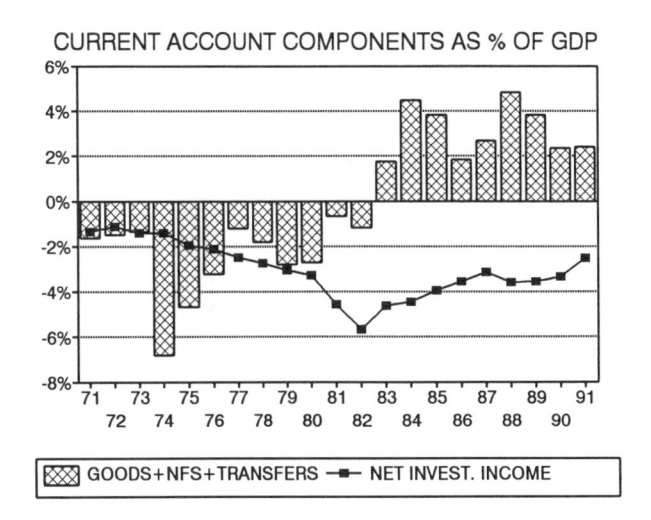

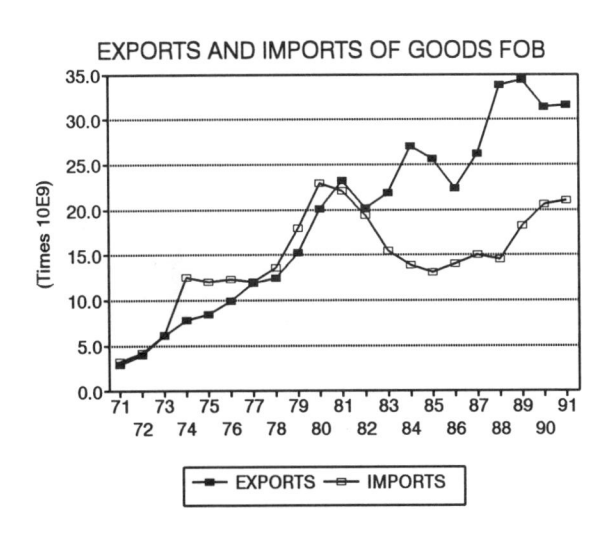

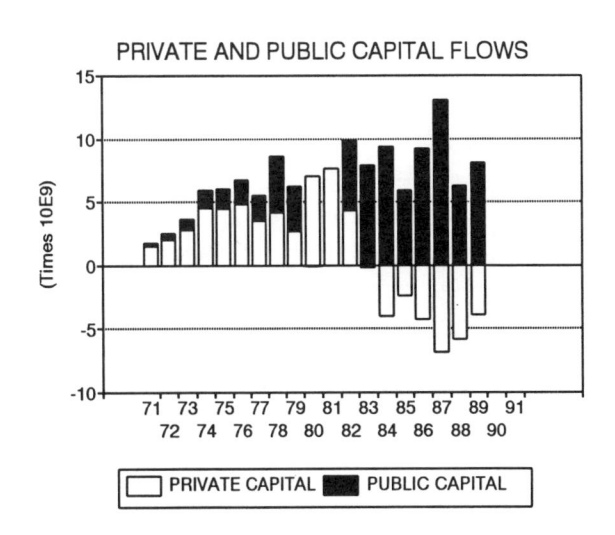

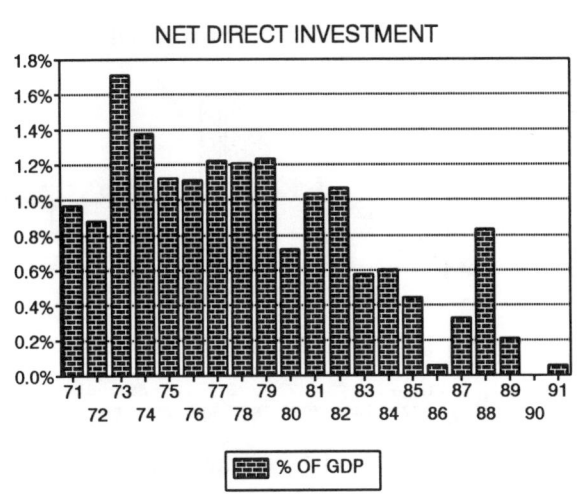

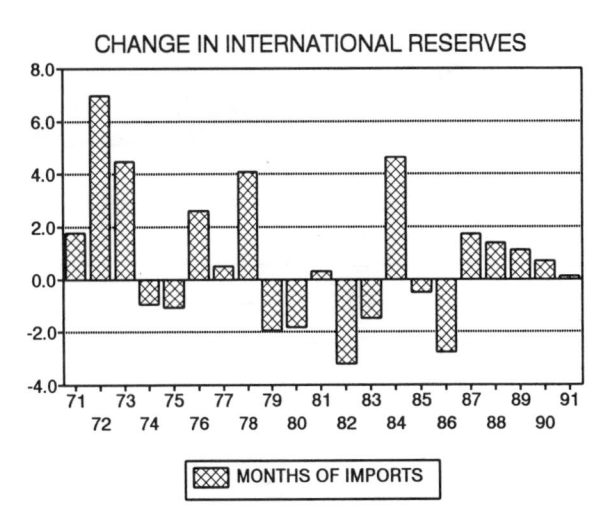

Statistics and Quantitative Analysis/IDB

43

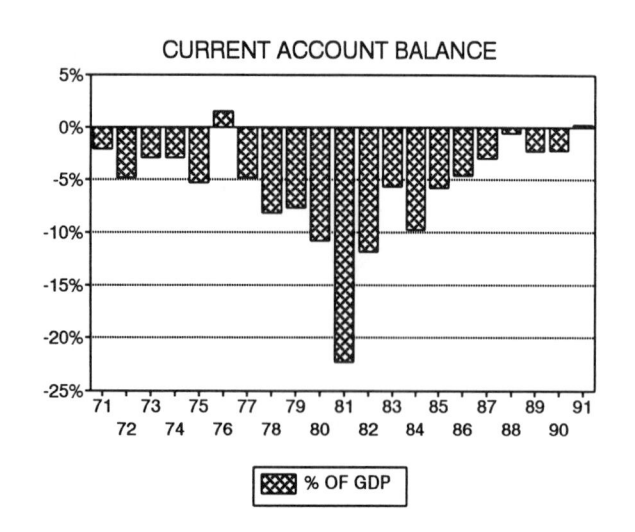

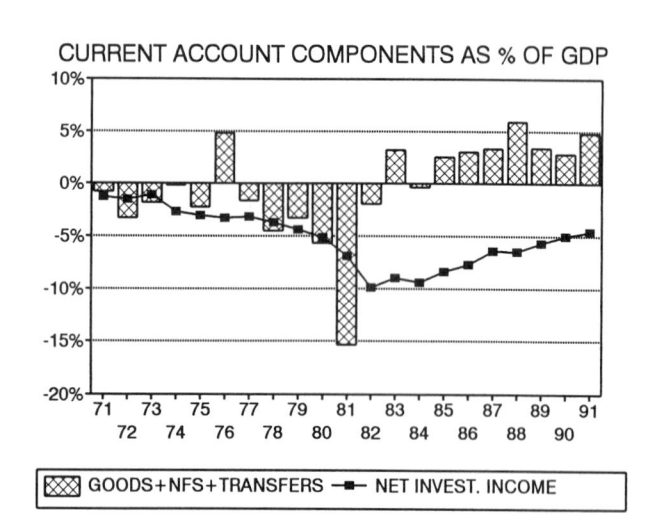

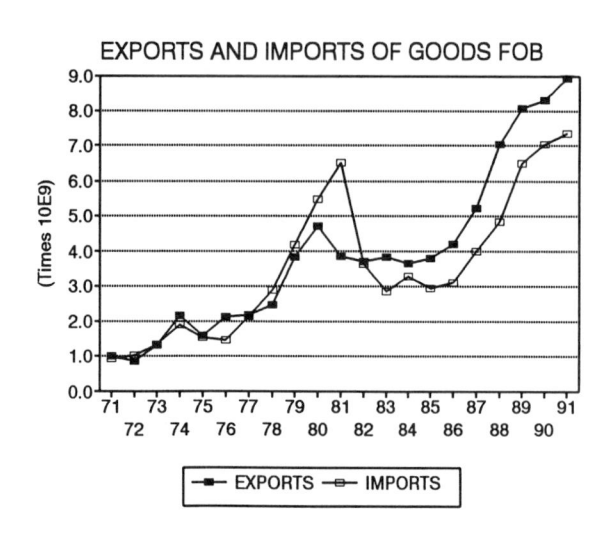

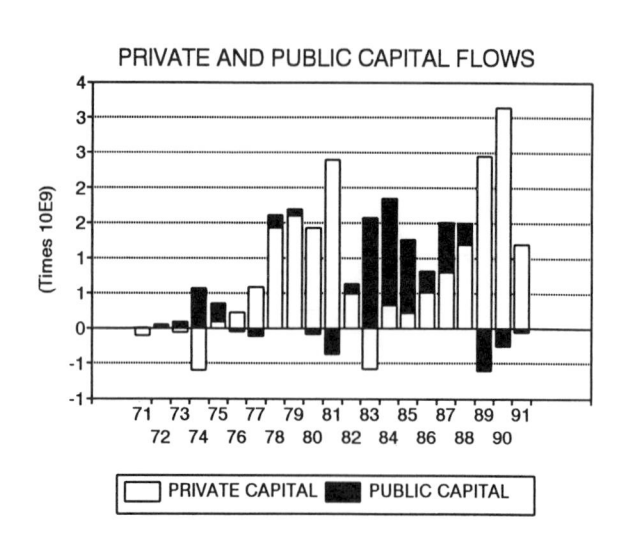

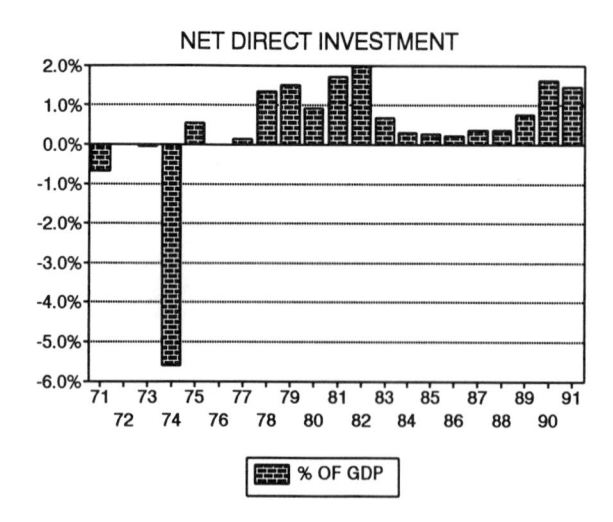

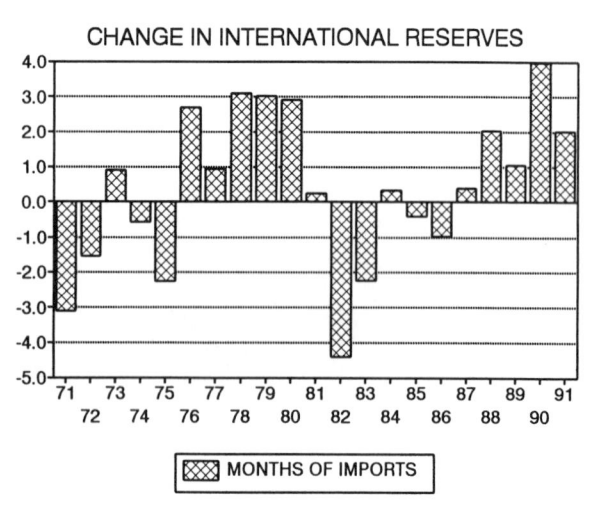

# CHILE
## Balance of Payments in US Dollars

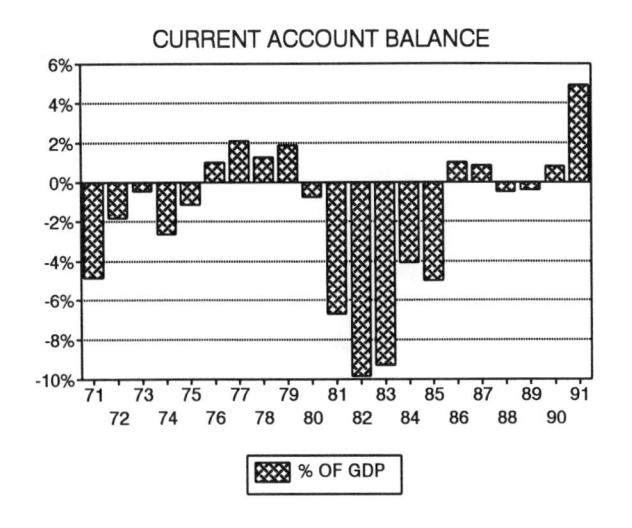

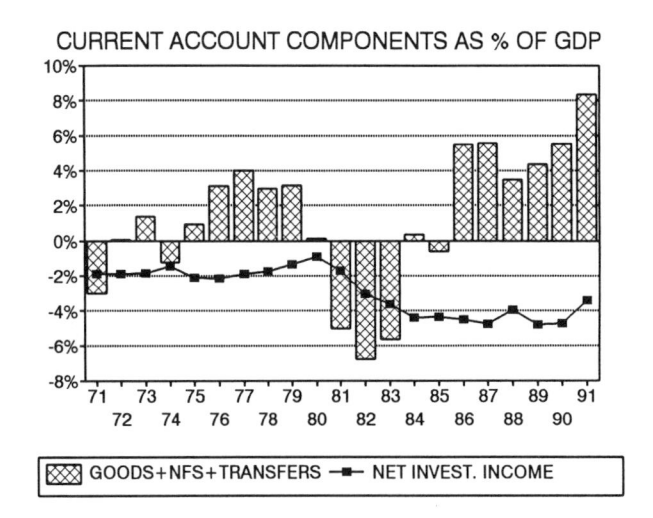

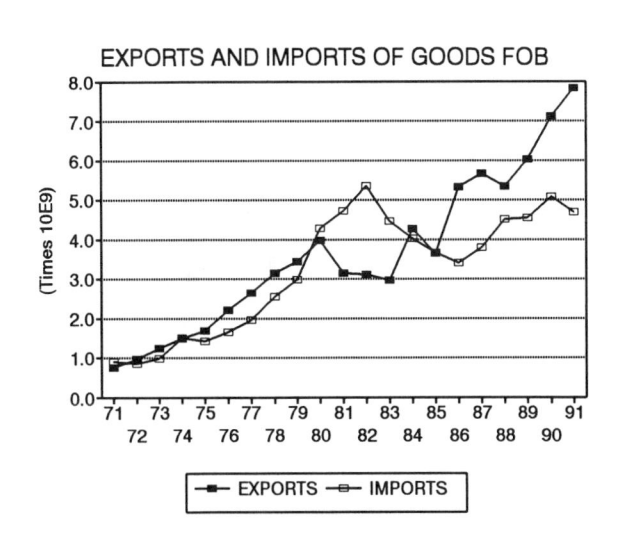

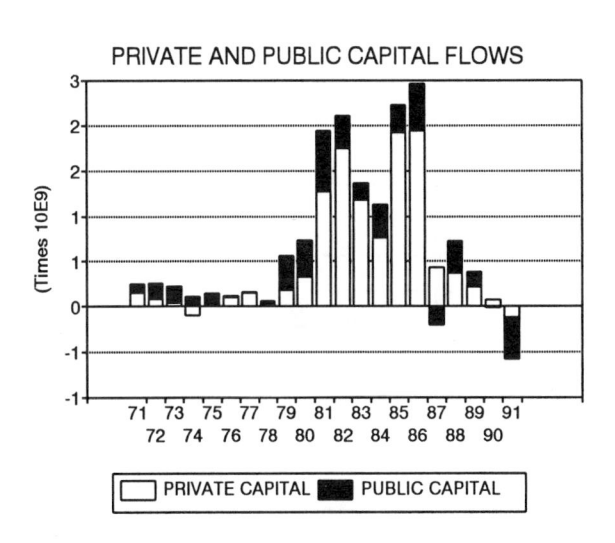

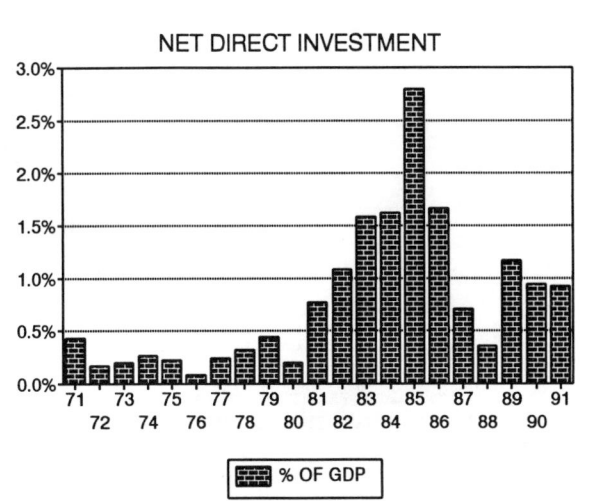

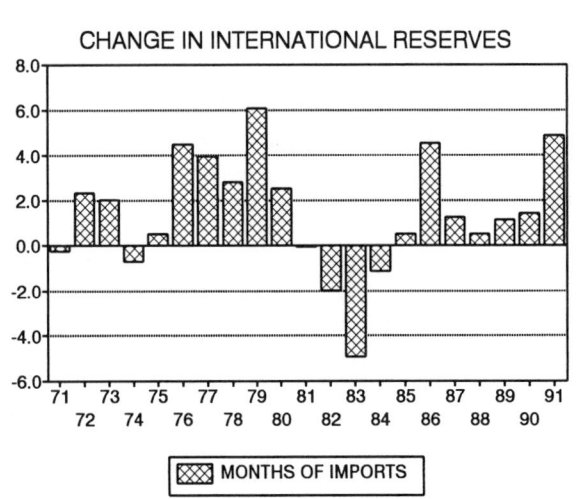

# COLOMBIA
## Balance of Payments in US Dollars

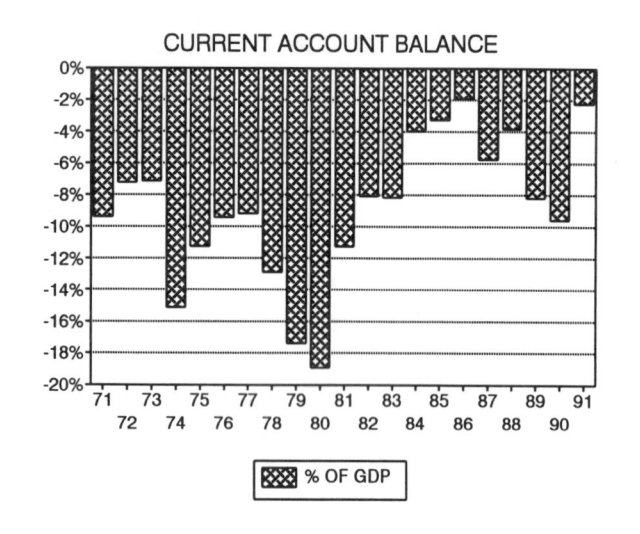

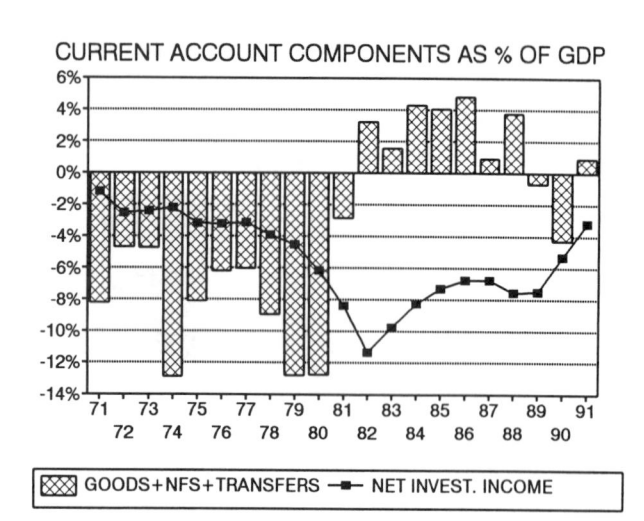

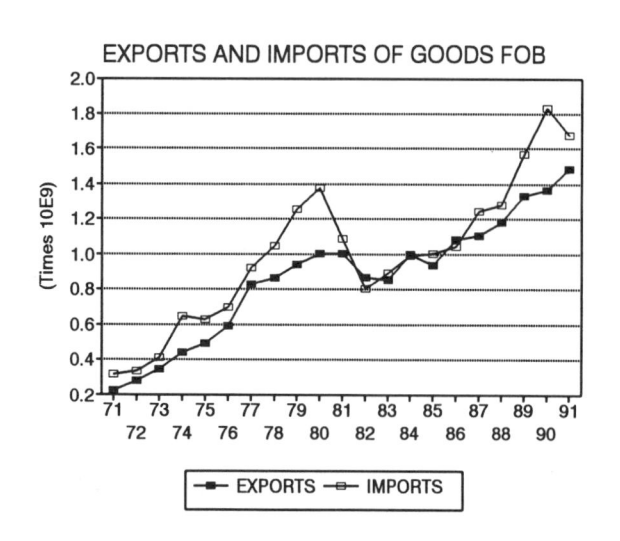

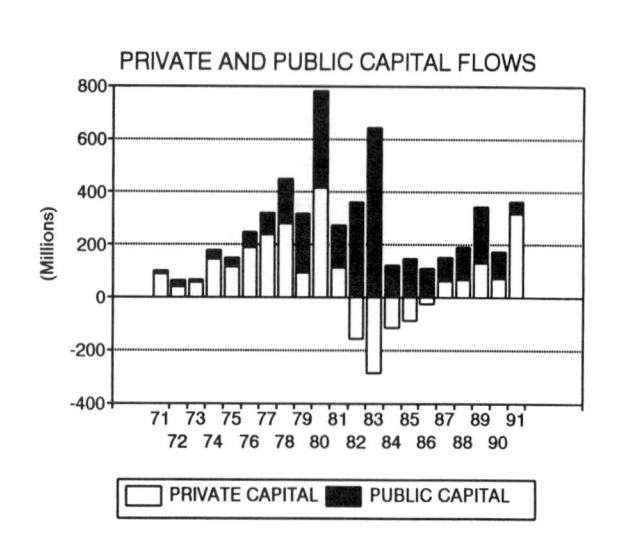

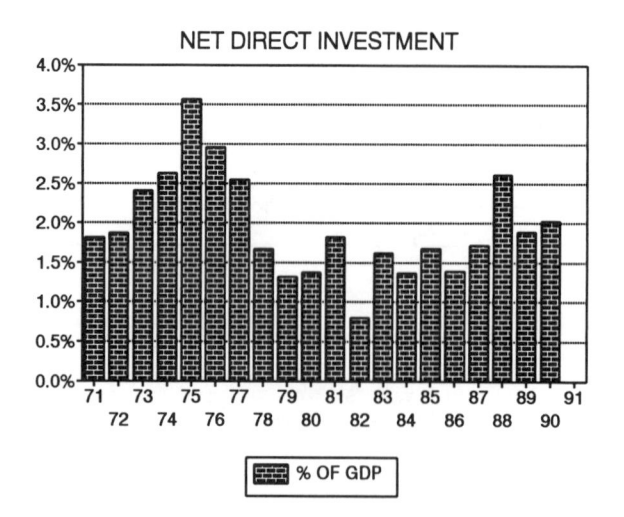

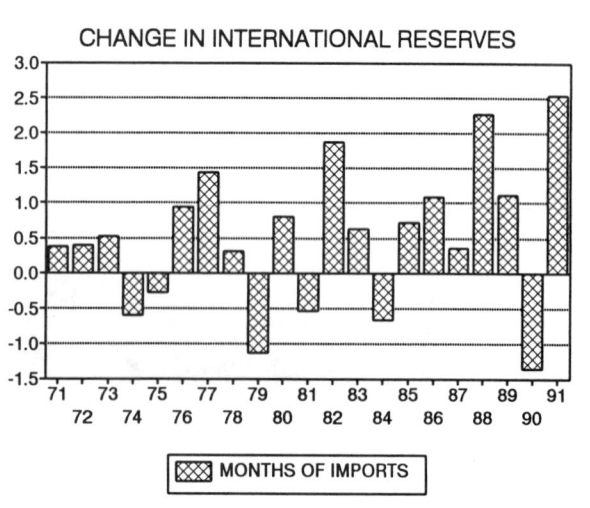

# COSTA RICA
## Balance of Payments in US Dollars

Statistics and Quantitative Analysis/IDB

# DOMINICAN REPUBLIC
## Balance of Payments in US Dollars

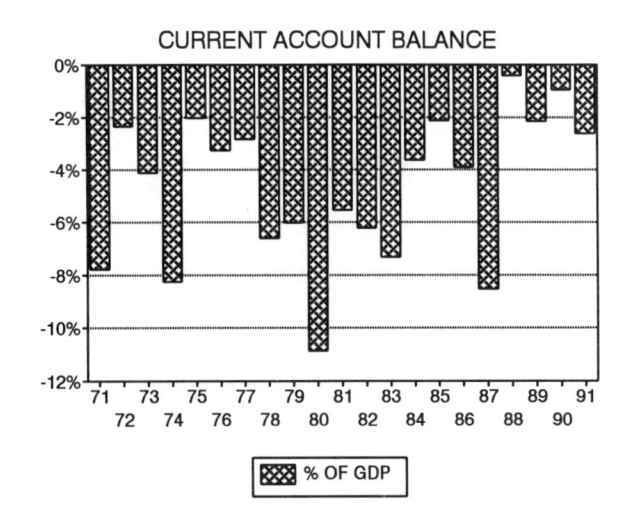

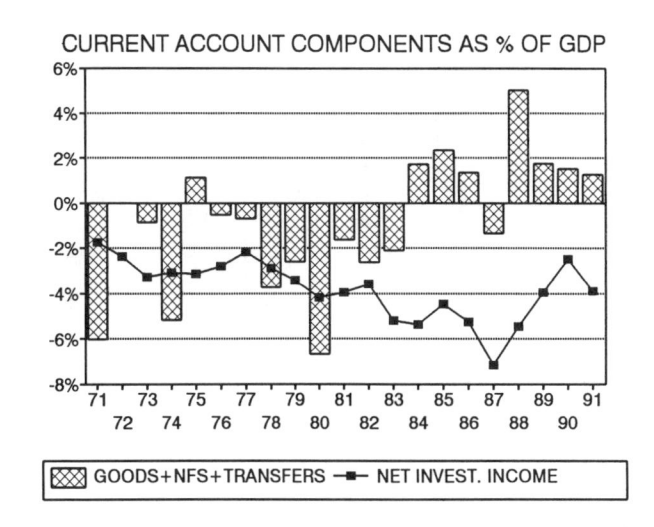

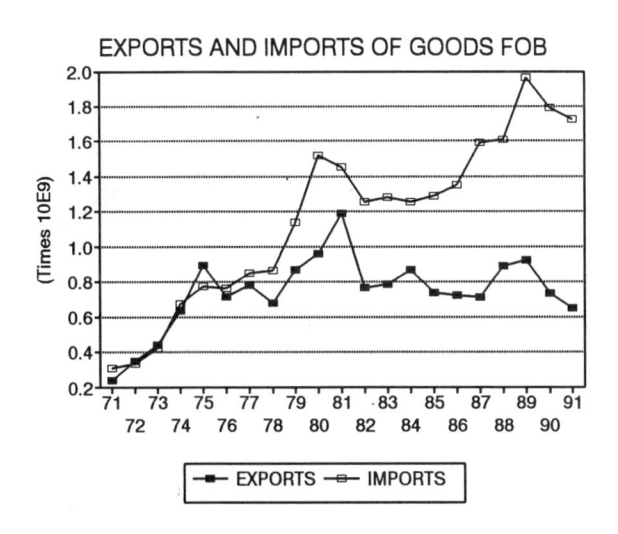

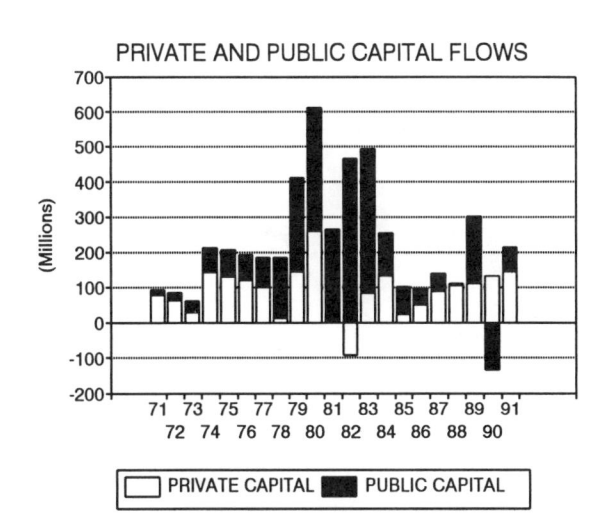

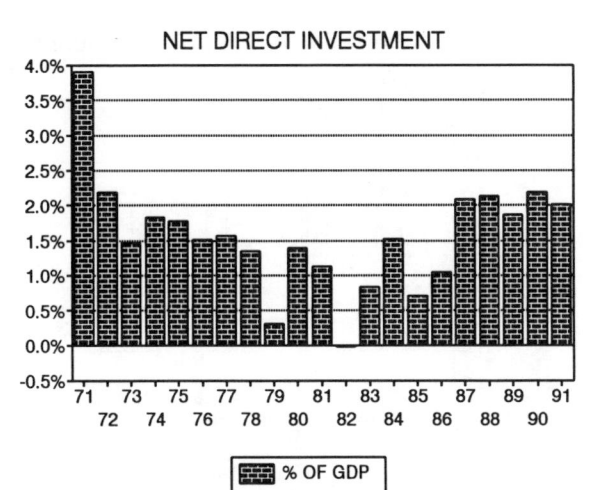

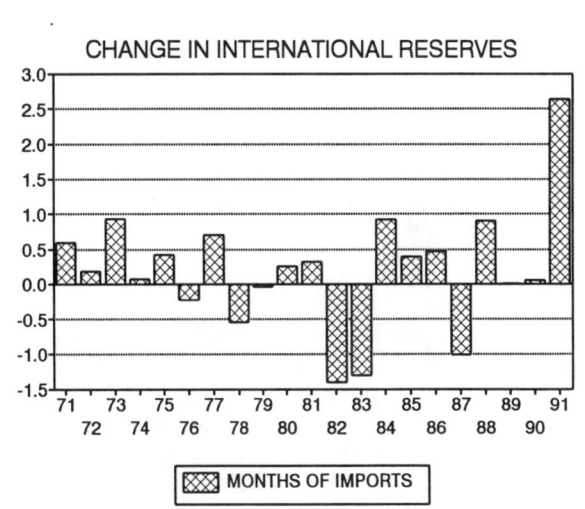

Statistics and Quantitative Analysis/IDB

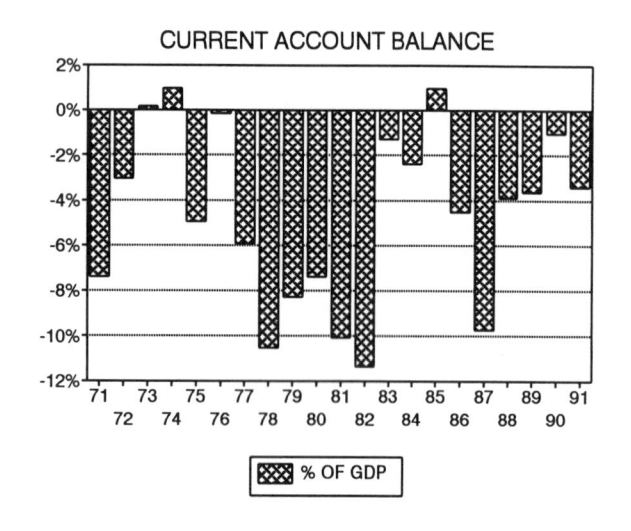

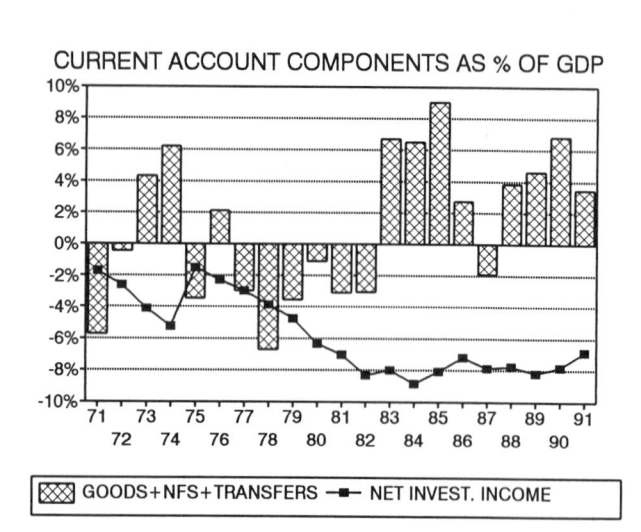

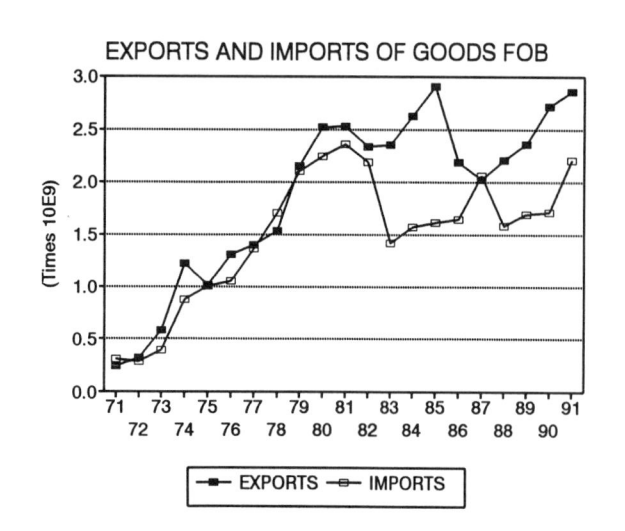

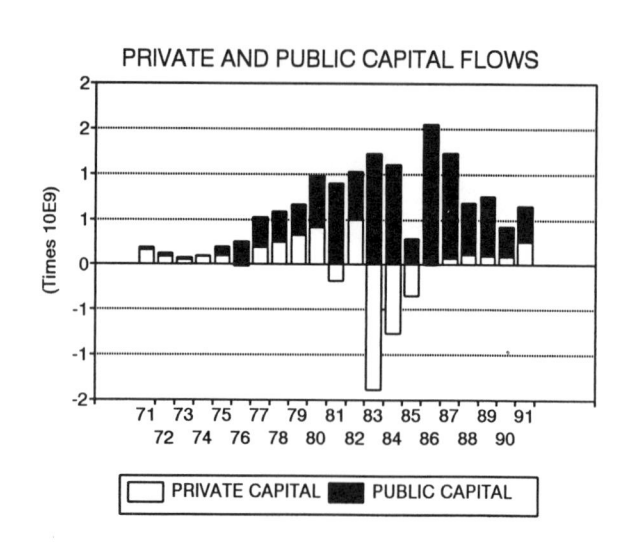

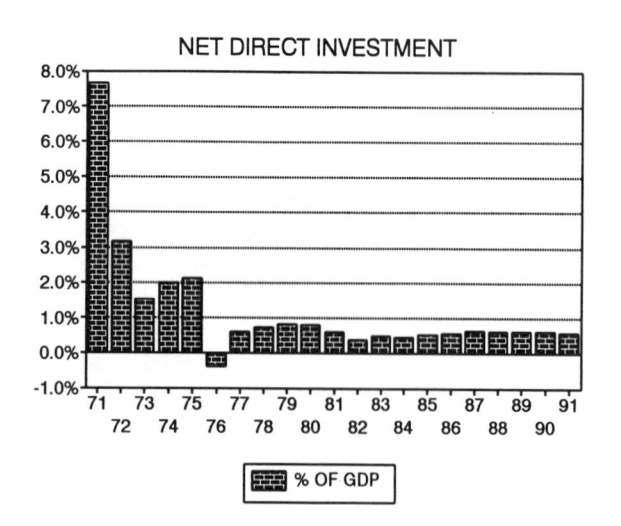

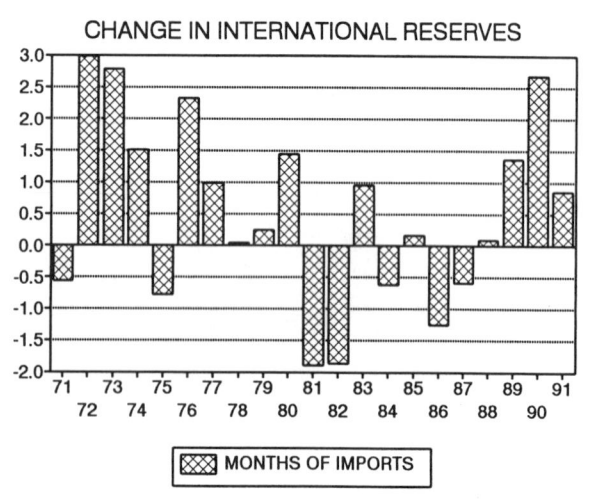

# ECUADOR
## Balance of Payments in US Dollars

# EL SALVADOR
## Balance of Payments in US Dollars

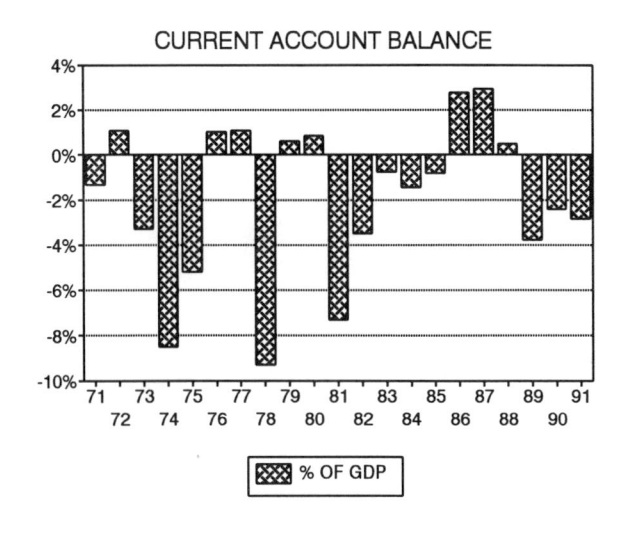

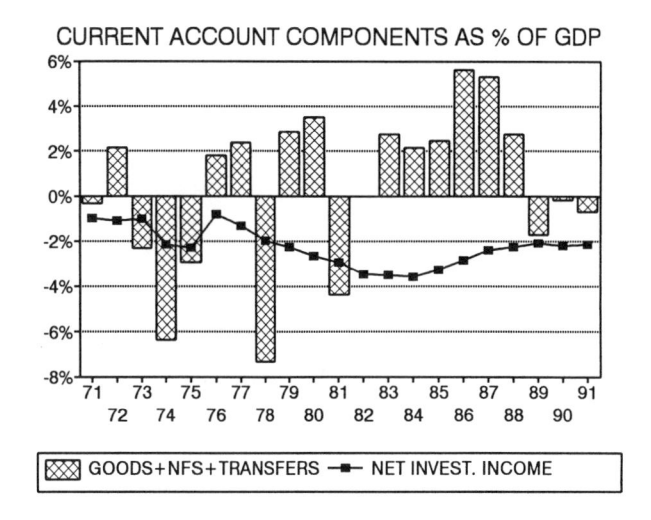

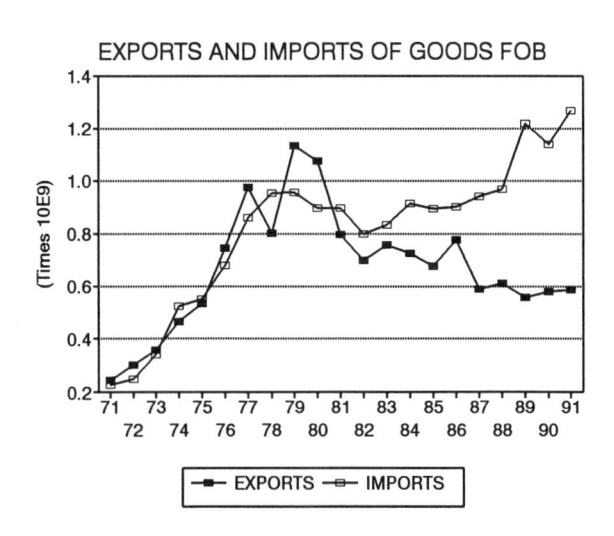

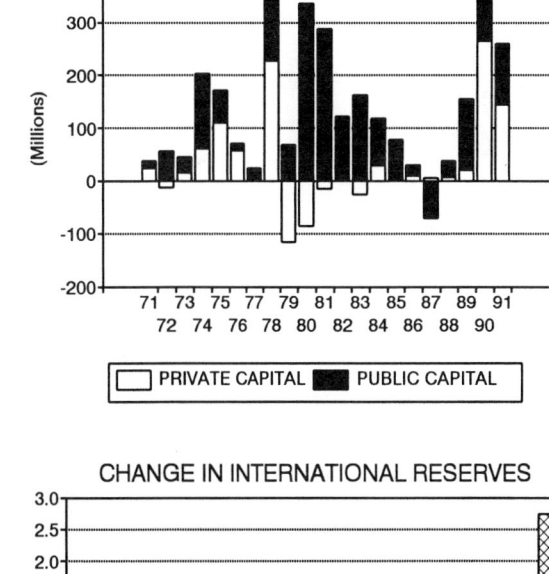

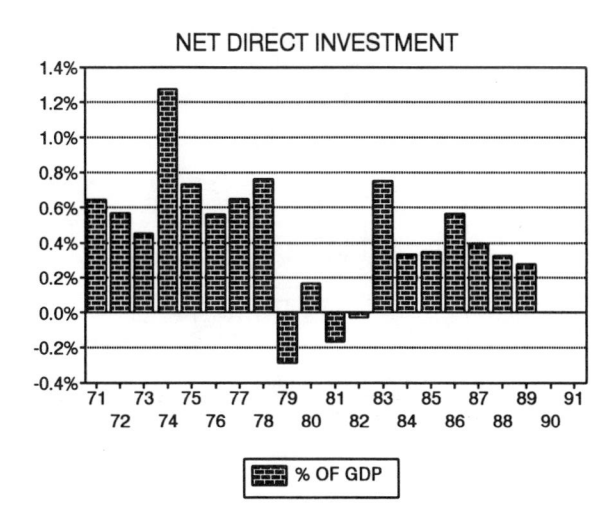

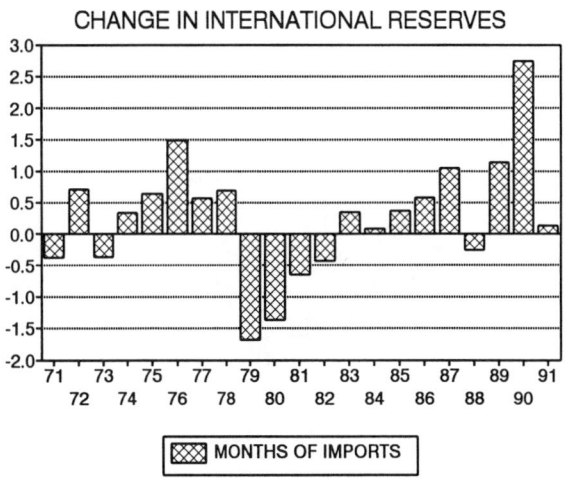

Statistics and Quantitative Analysis/IDB

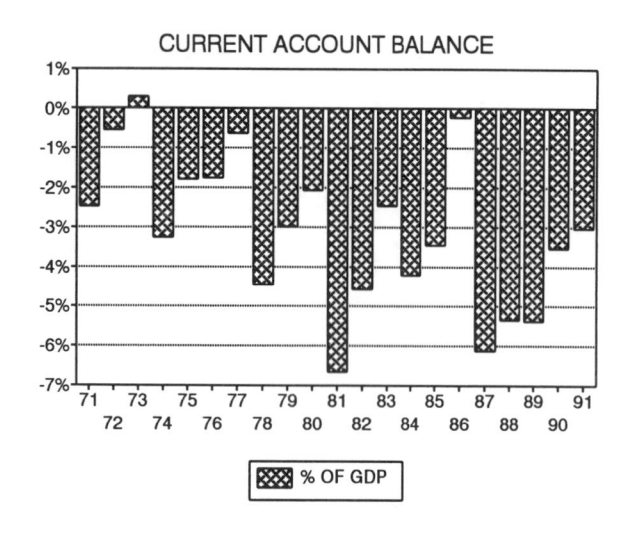

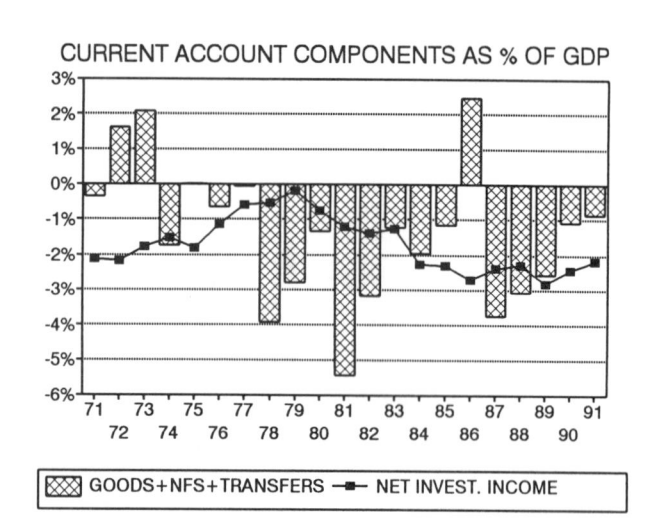

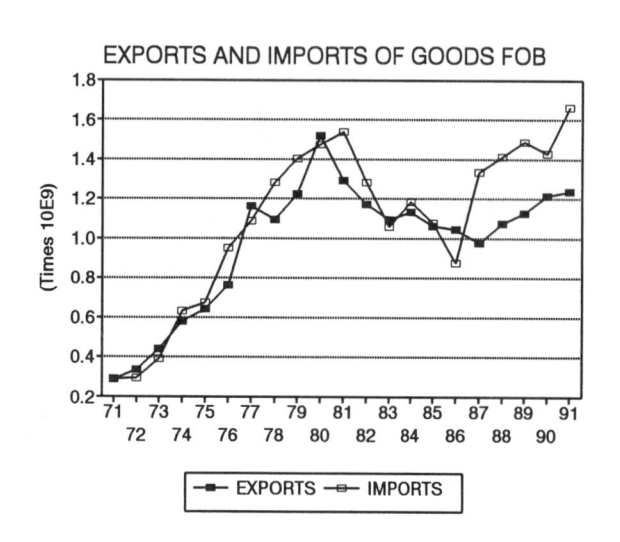

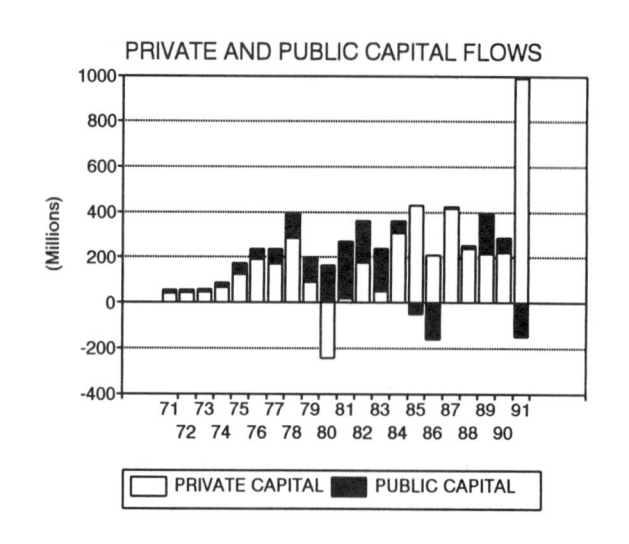

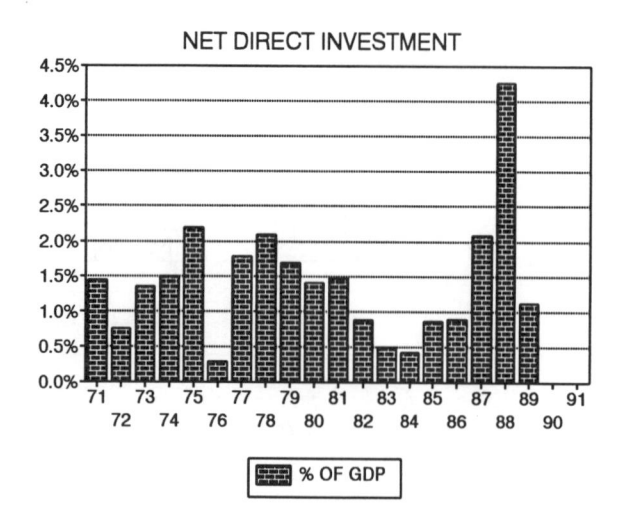

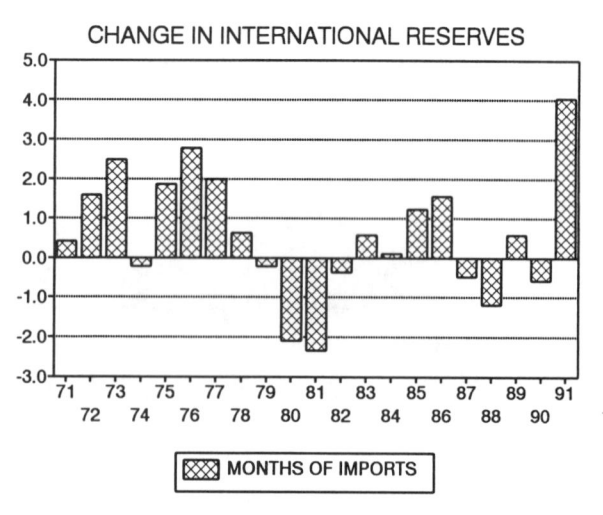

# GUATEMALA
## Balance of Payments in US Dollars

Statistics and Quantitative Analysis/IDB

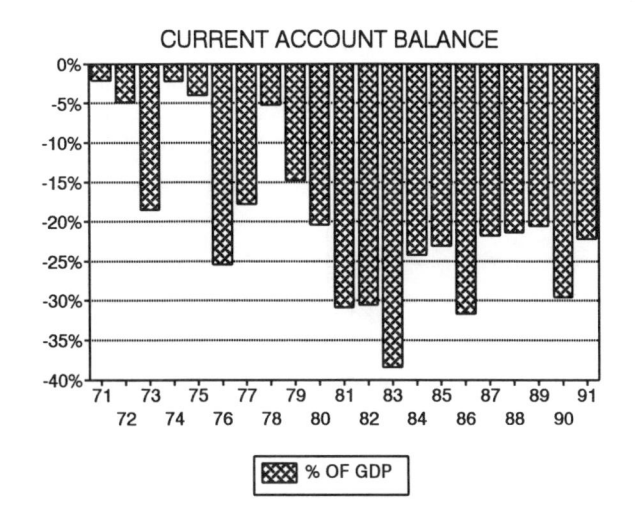

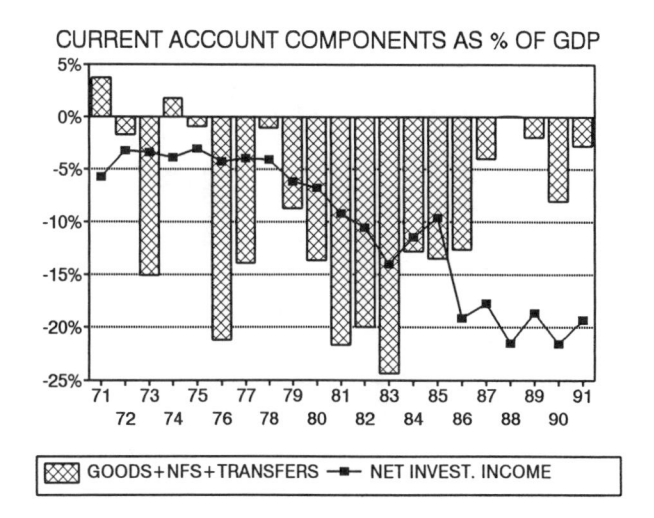

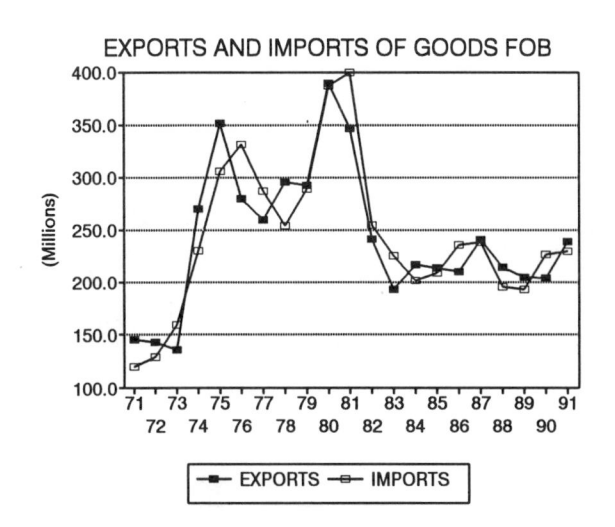

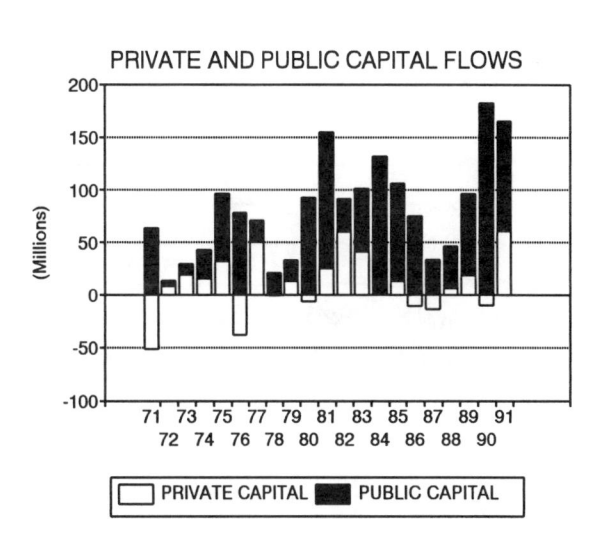

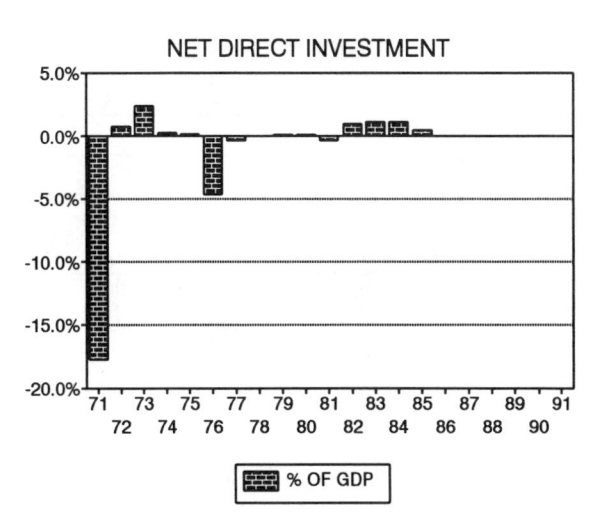

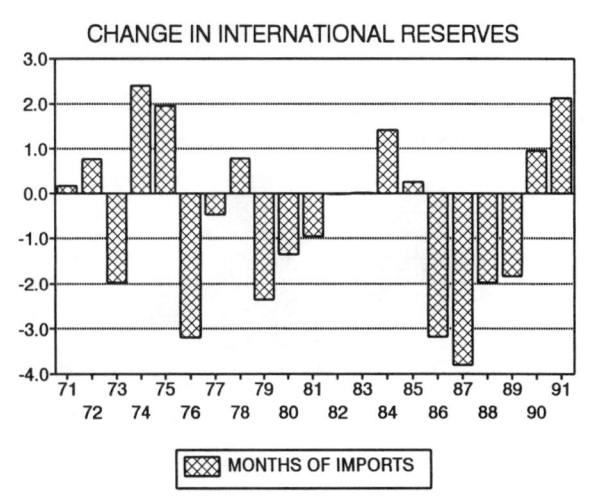

# GUYANA
## Balance of Payments in US Dollars

Statistics and Quantitative Analysis/IDB

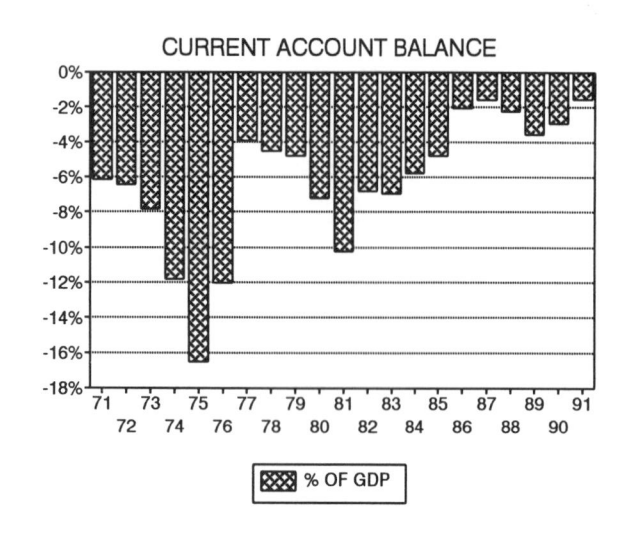

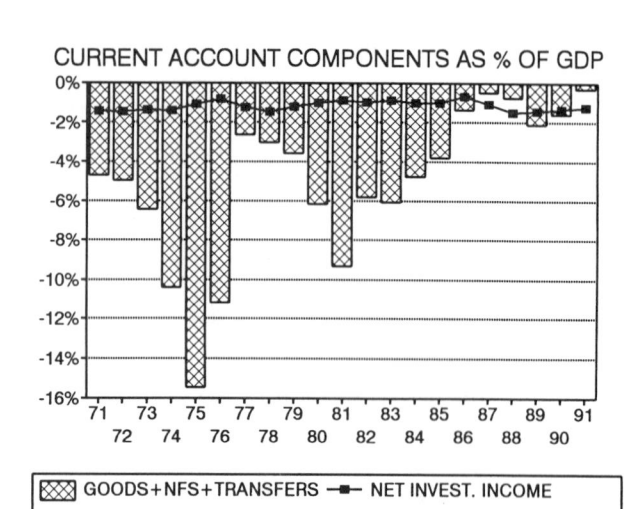

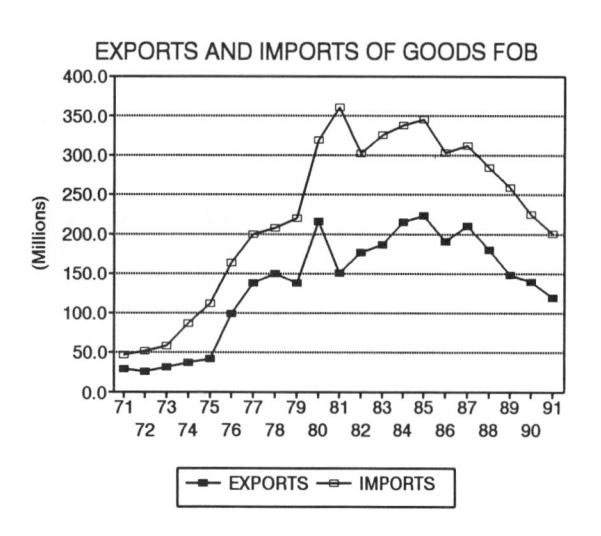

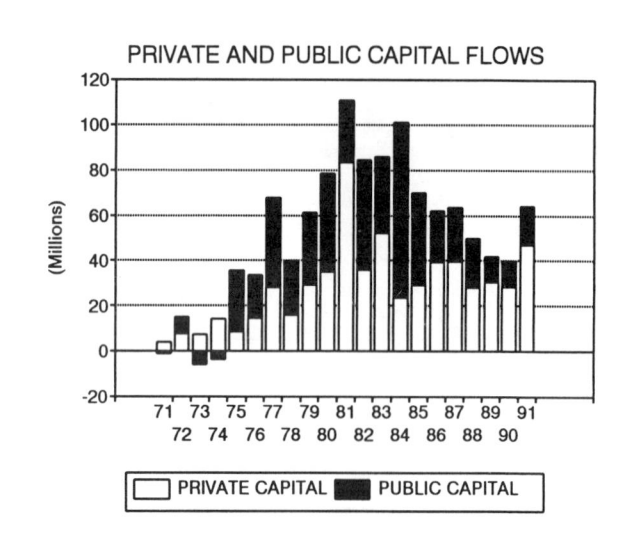

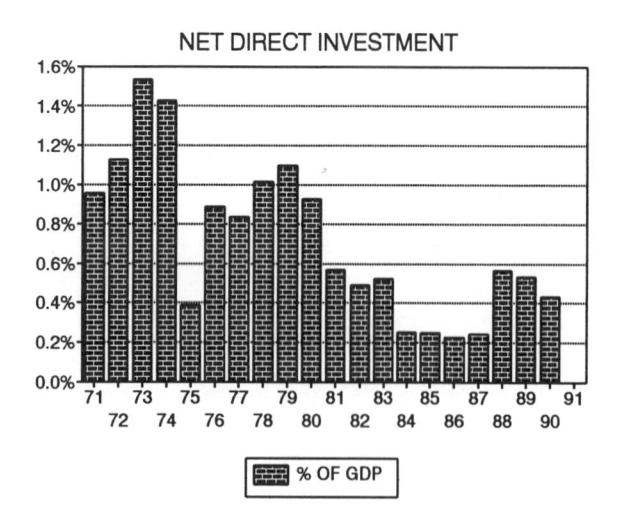

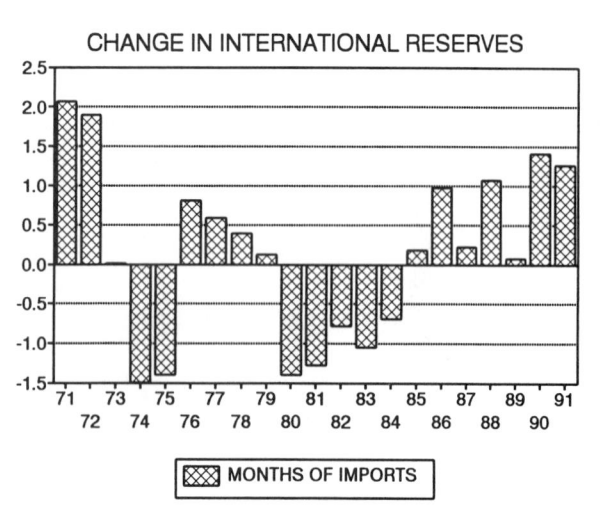

# HAITI
## Balance of Payments in US Dollars

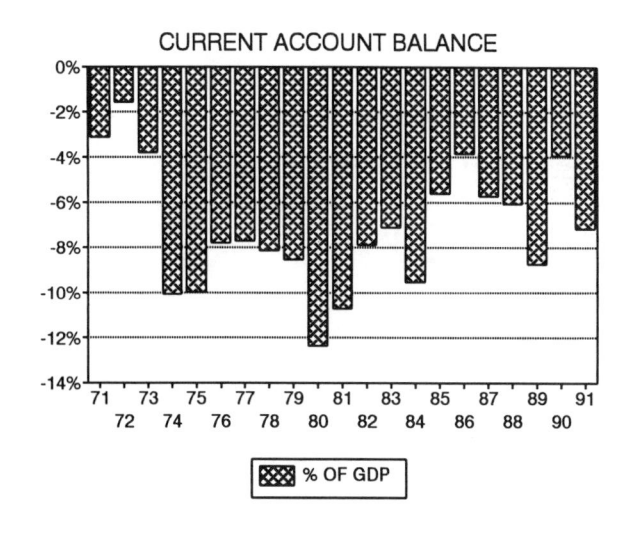

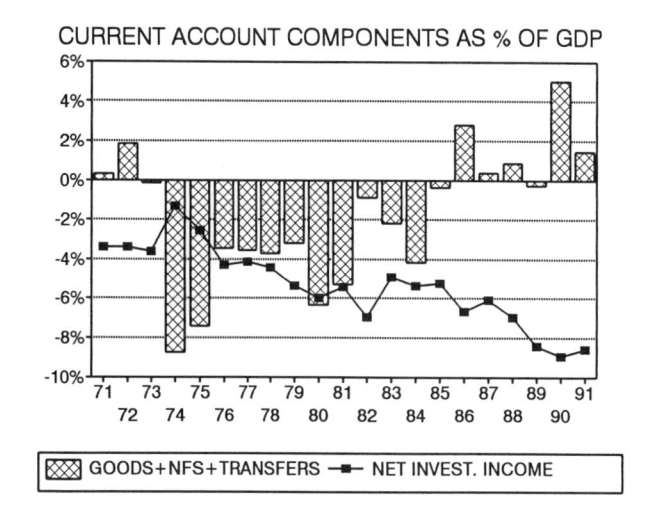

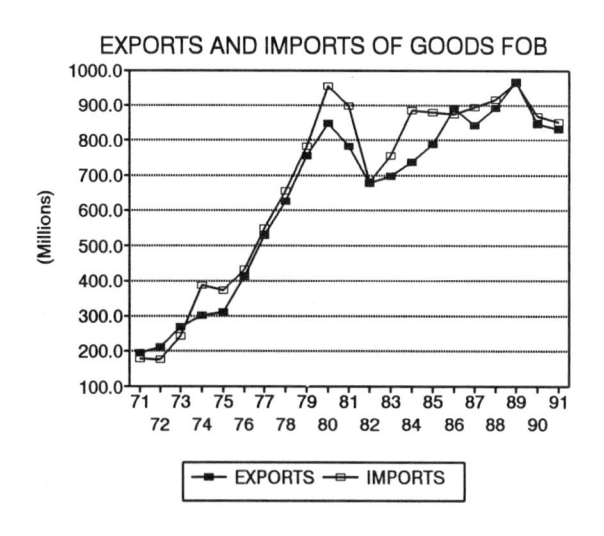

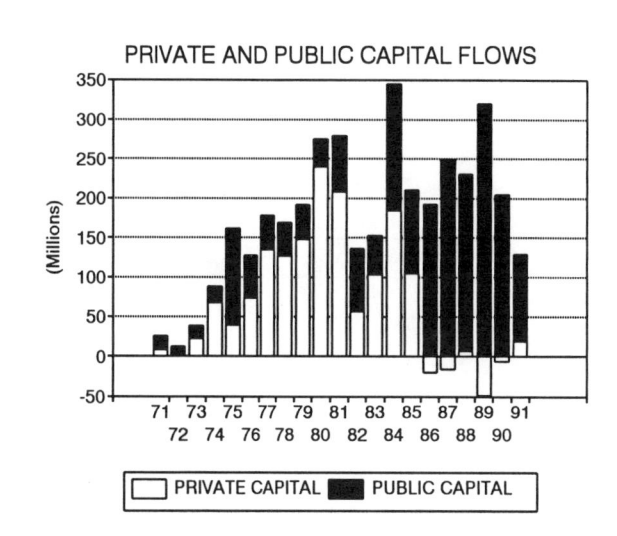

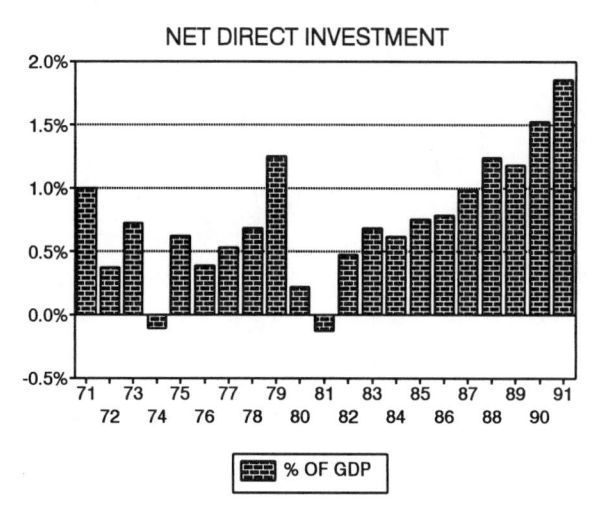

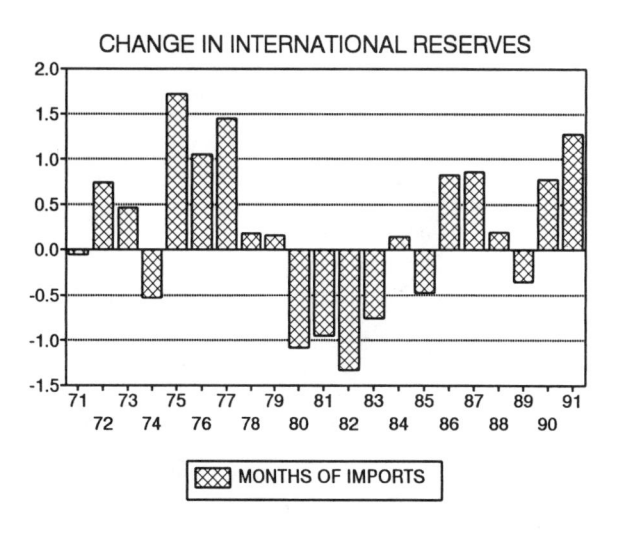

# HONDURAS
## Balance of Payments in US Dollars

Statistics and Quantitative Analysis/IDB

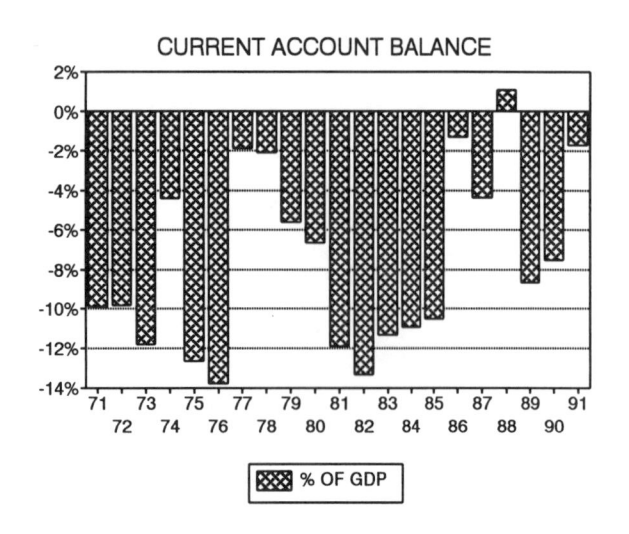

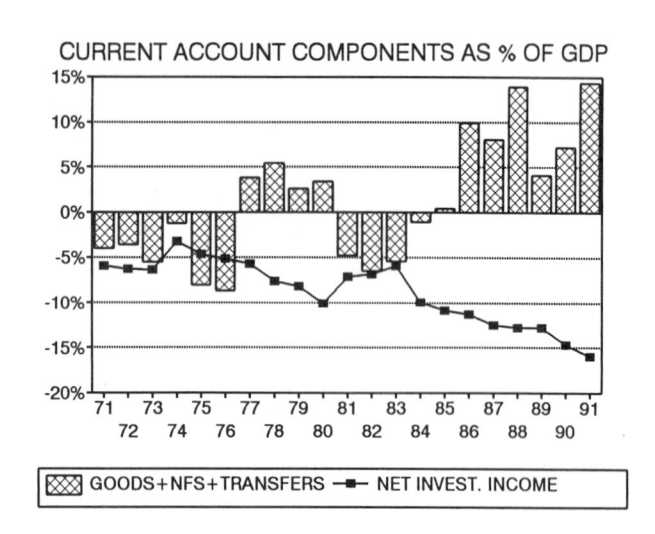

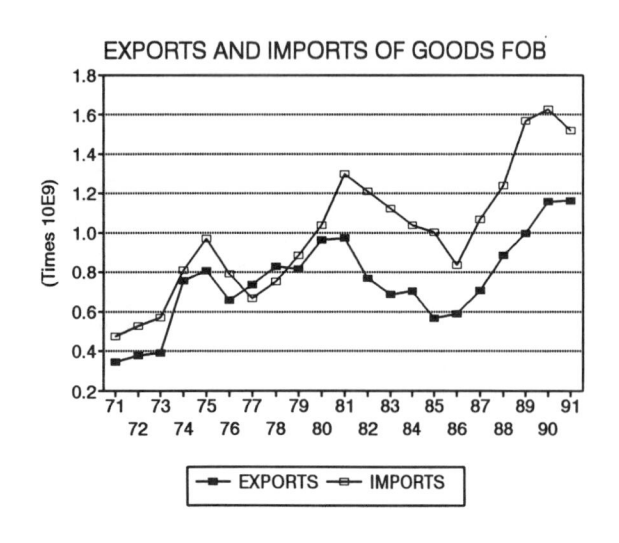

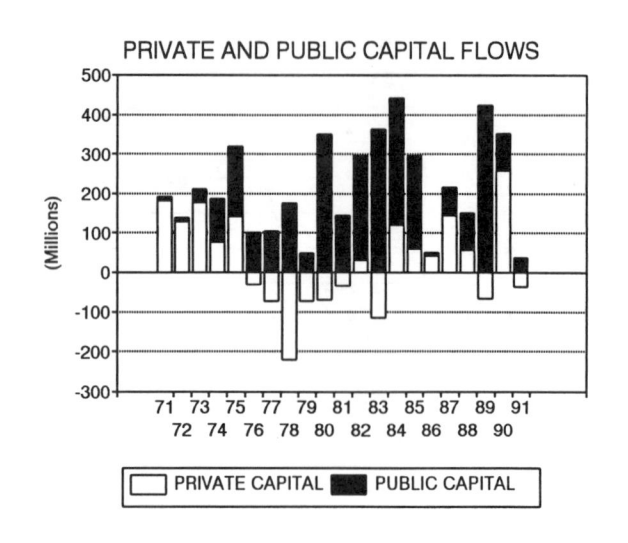

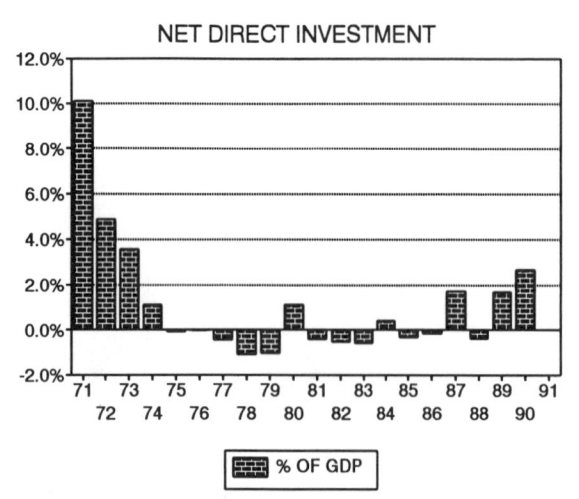

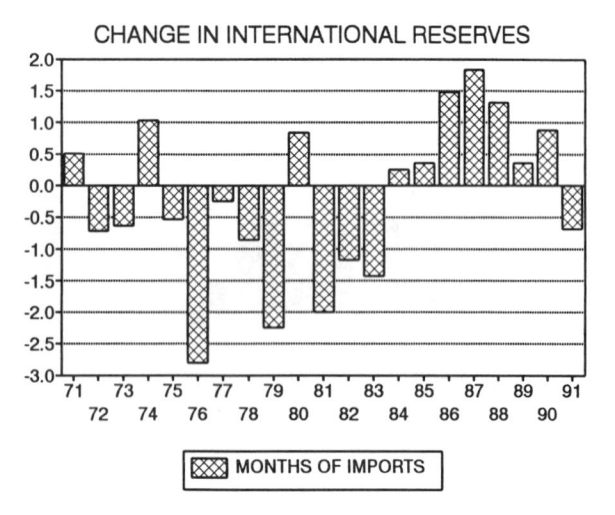

# JAMAICA
## Balance of Payments in US Dollars

Statistics and Quantitative Analysis/IDB

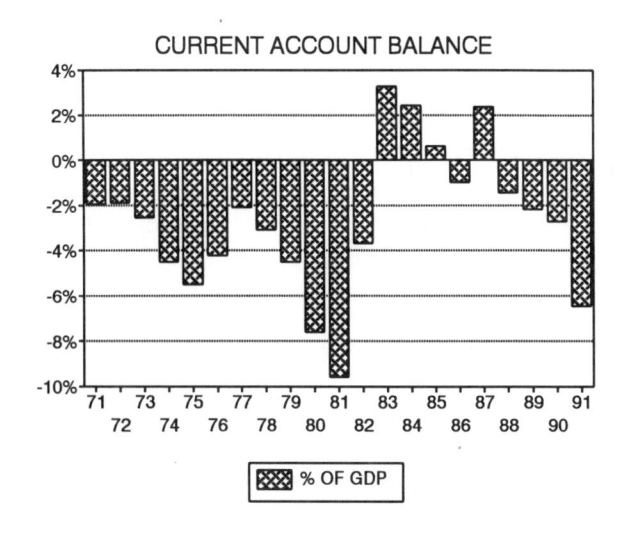

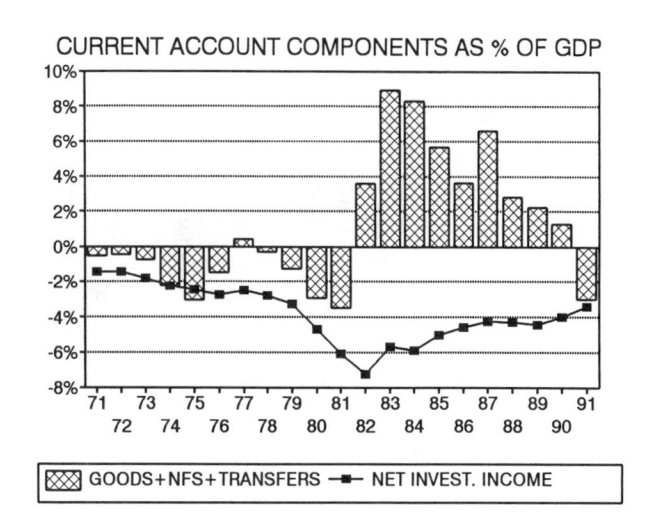

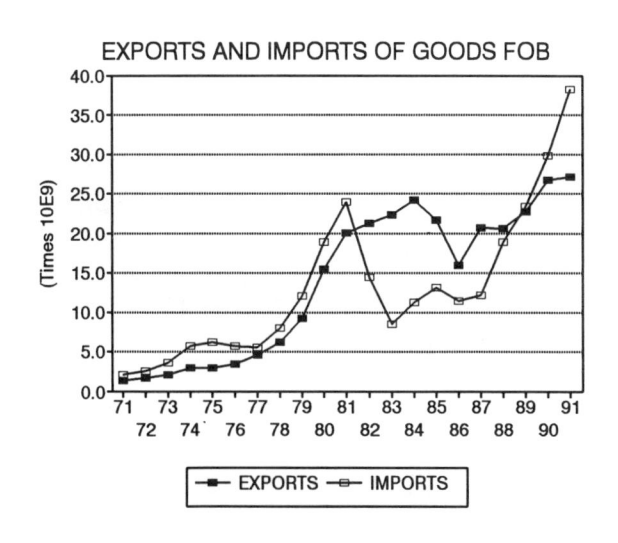

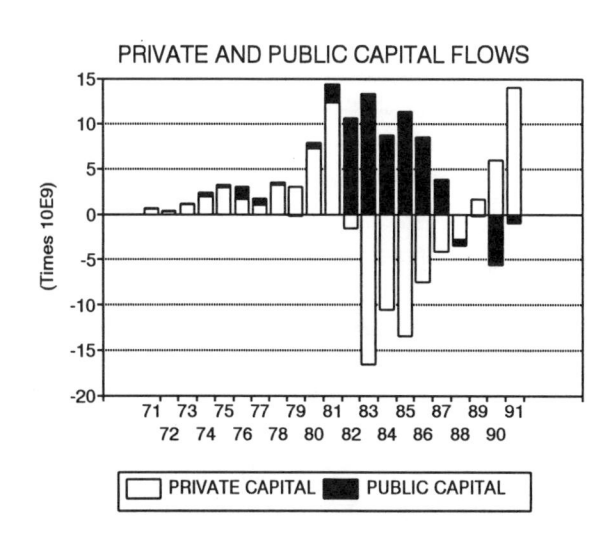

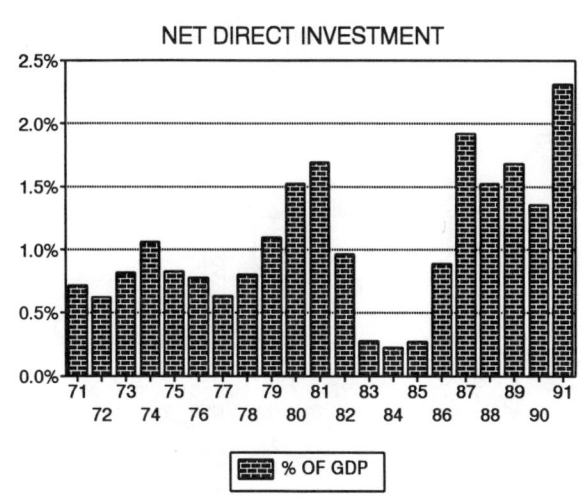

# MEXICO
## Balance of Payments in US Dollars

Statistics and Quantitative Analysis/IDB

55

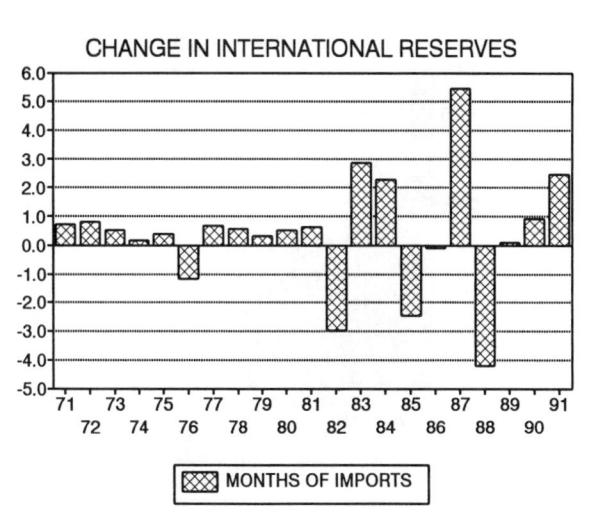

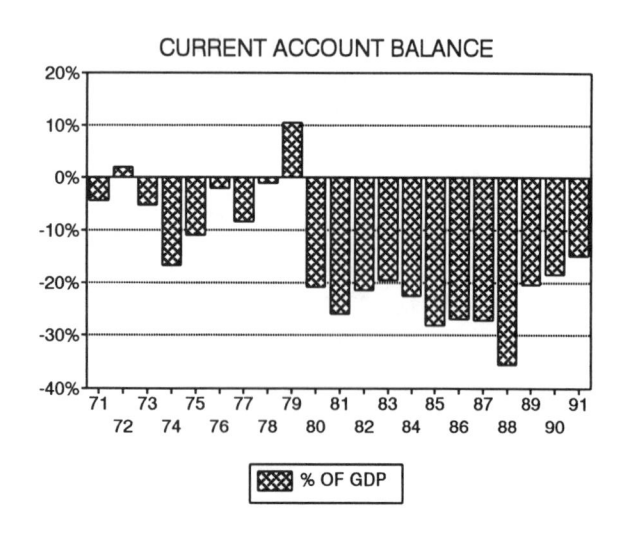

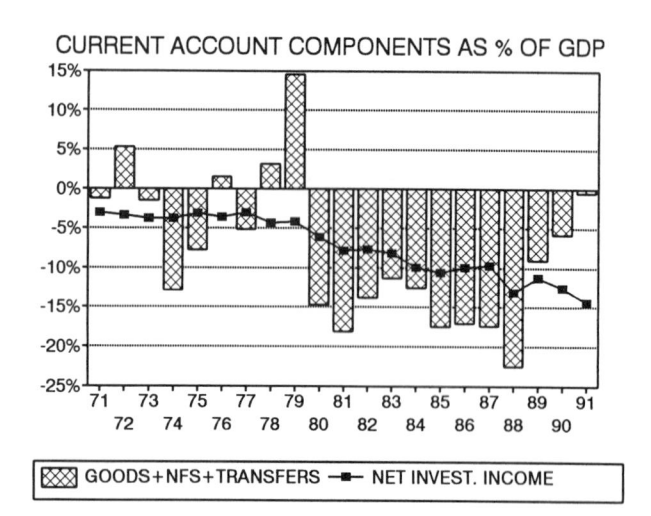

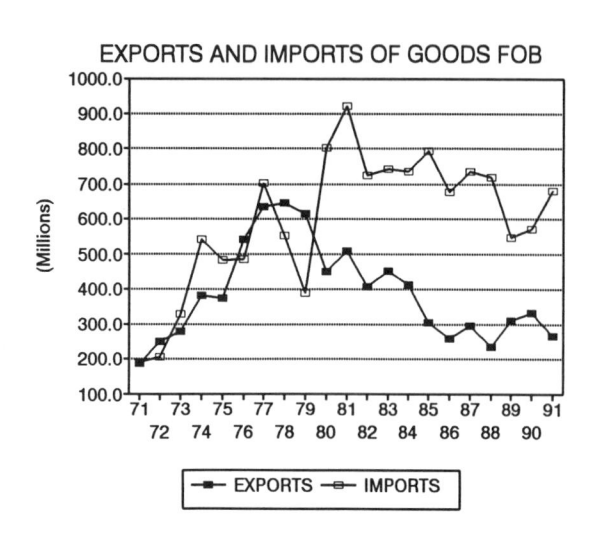

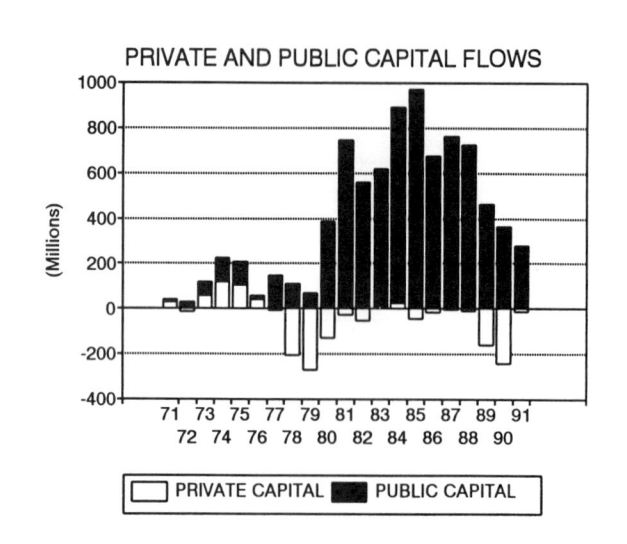

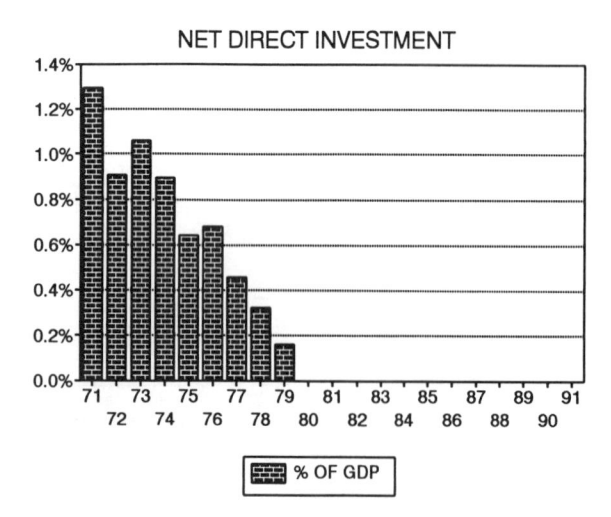

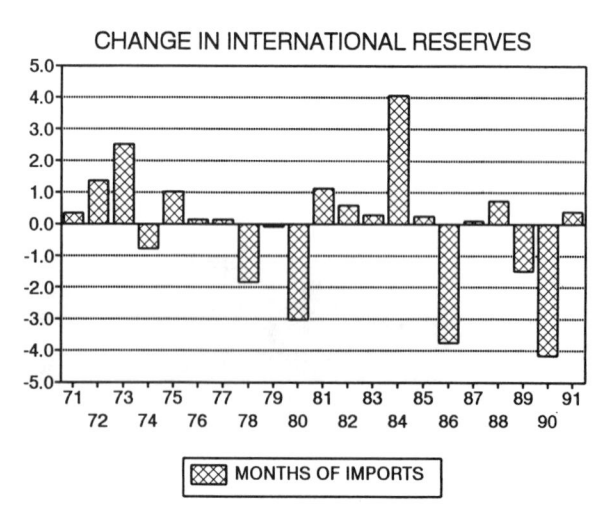

# NICARAGUA
## Balance of Payments in US Dollars

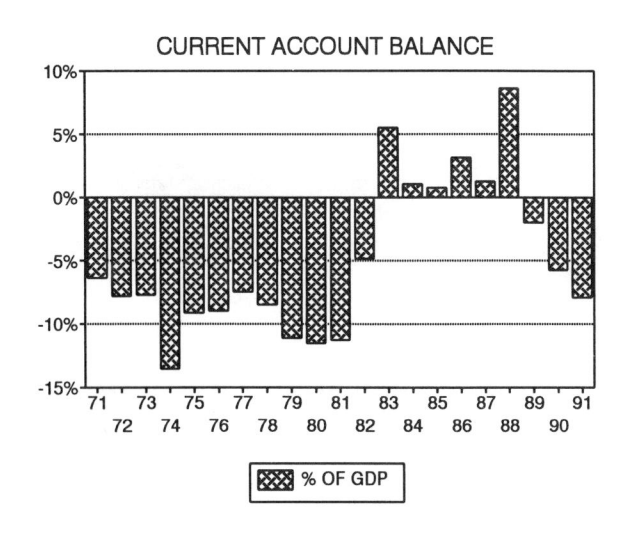

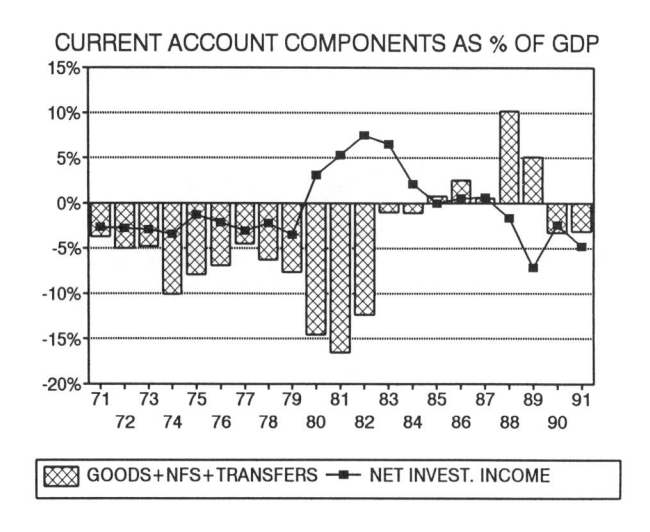

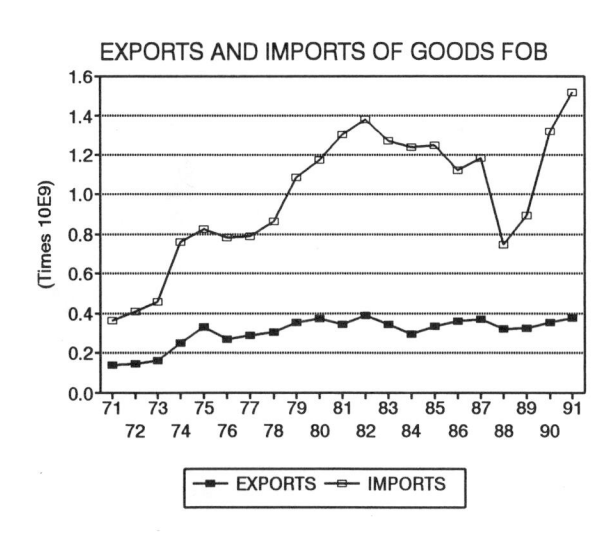

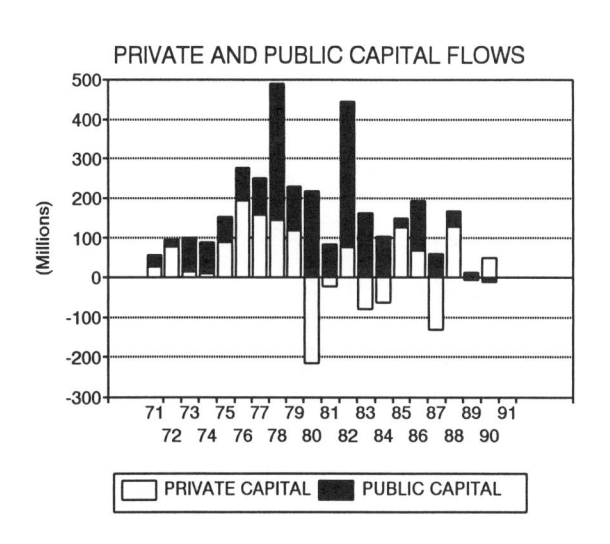

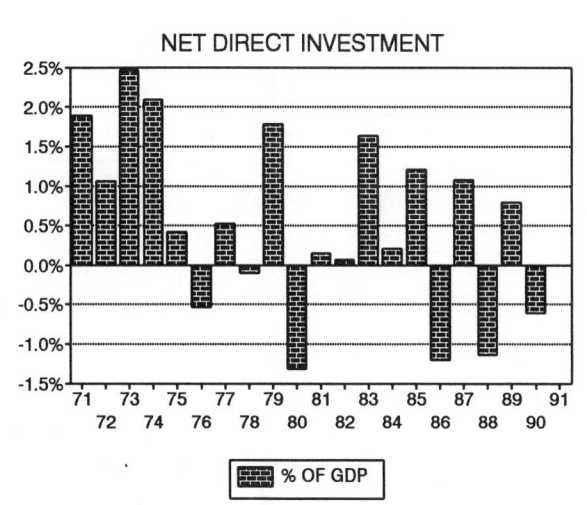

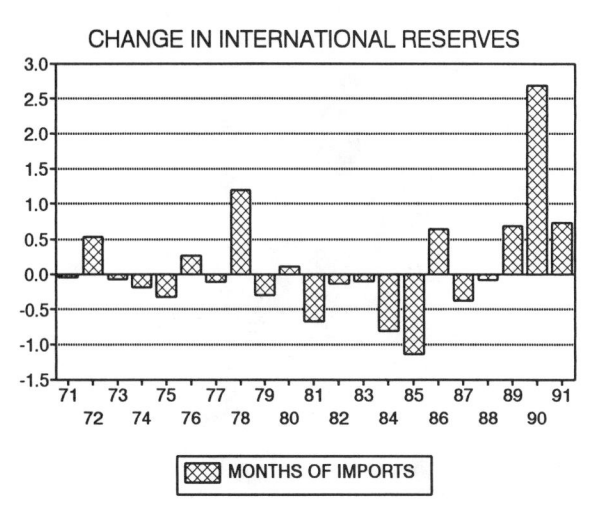

# PANAMA
## Balance of Payments in US Dollars

Statistics and Quantitative Analysis/IDB

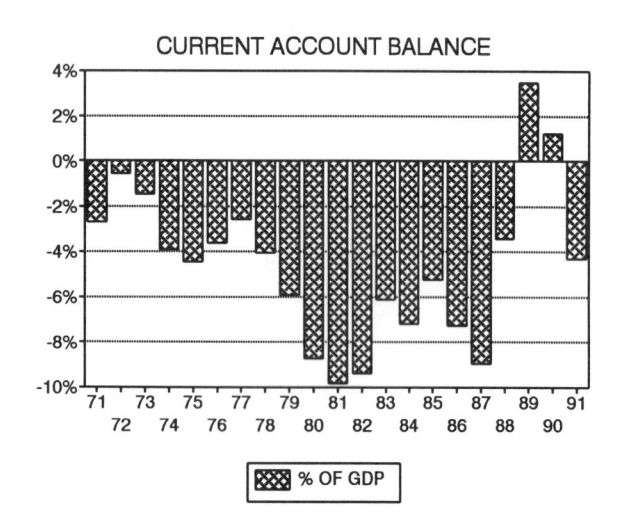

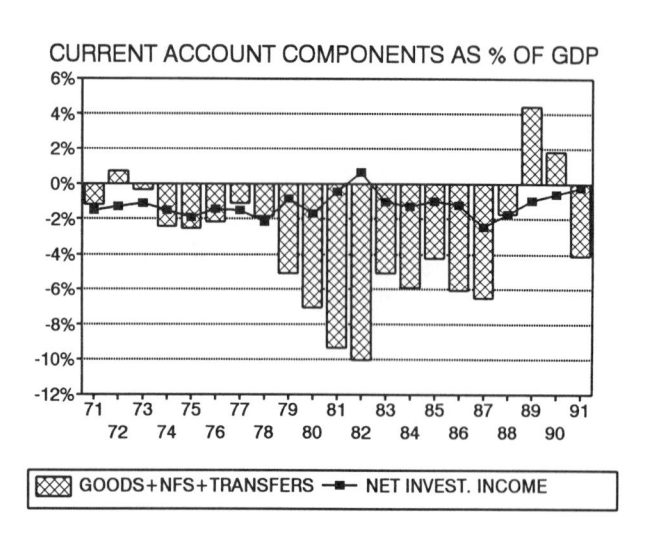

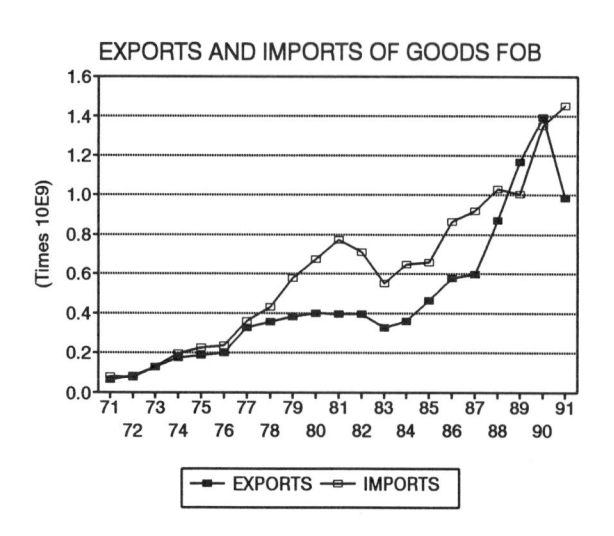

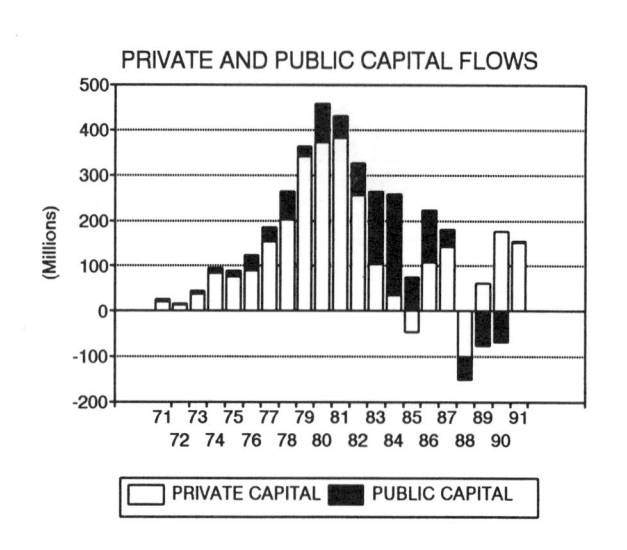

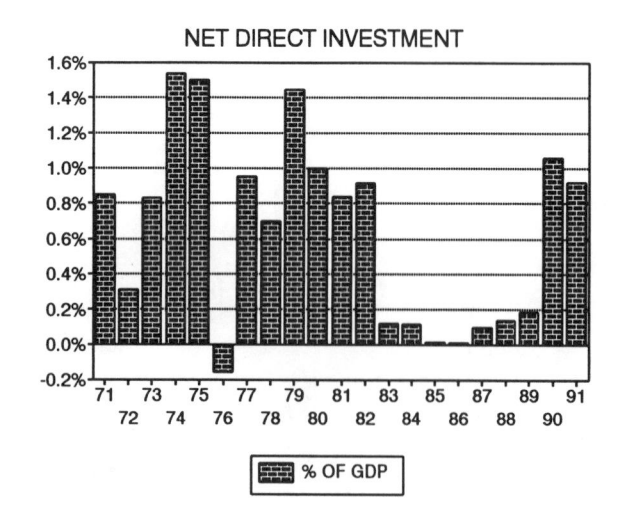

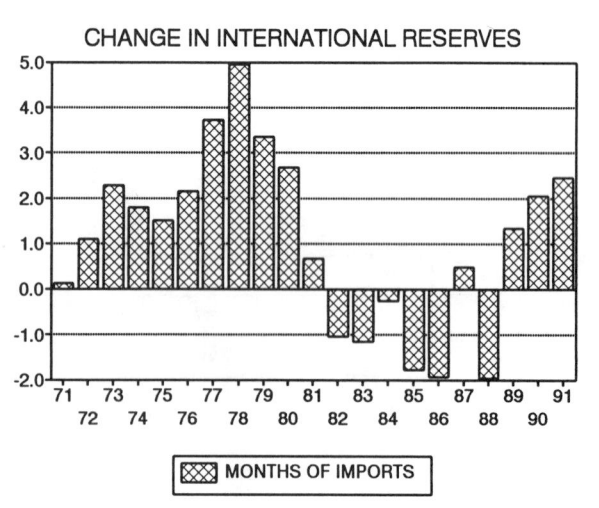

# PARAGUAY
## Balance of Payments in US Dollars

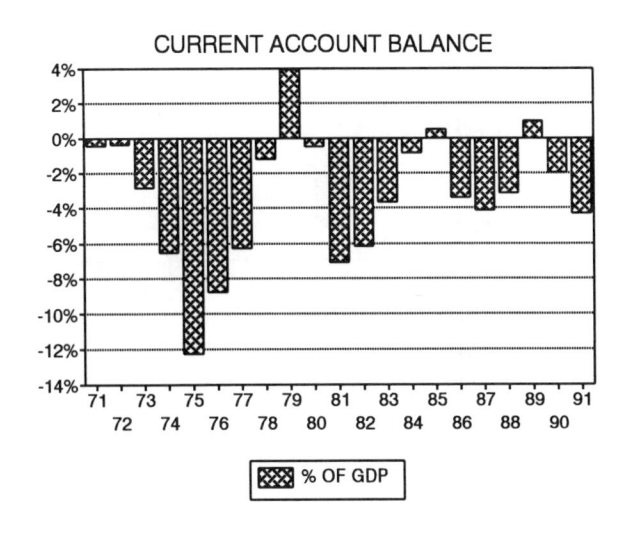

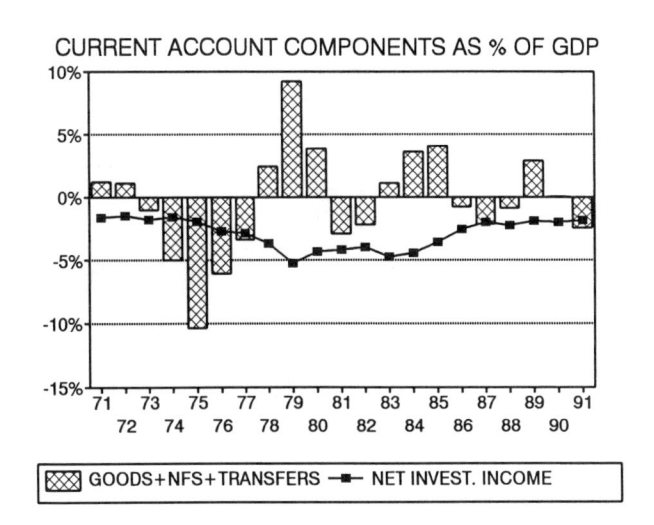

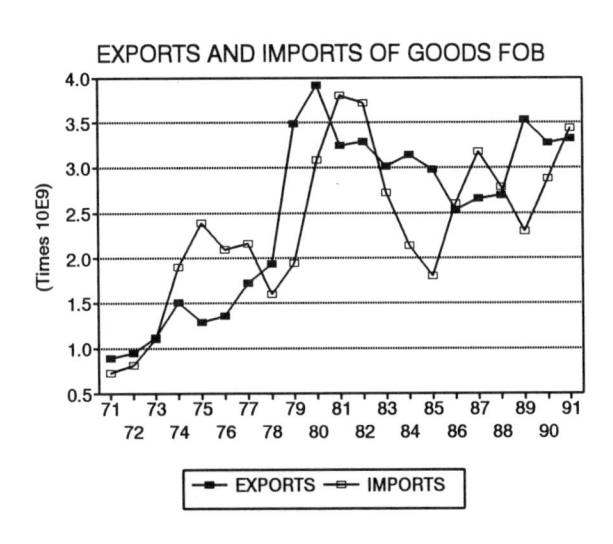

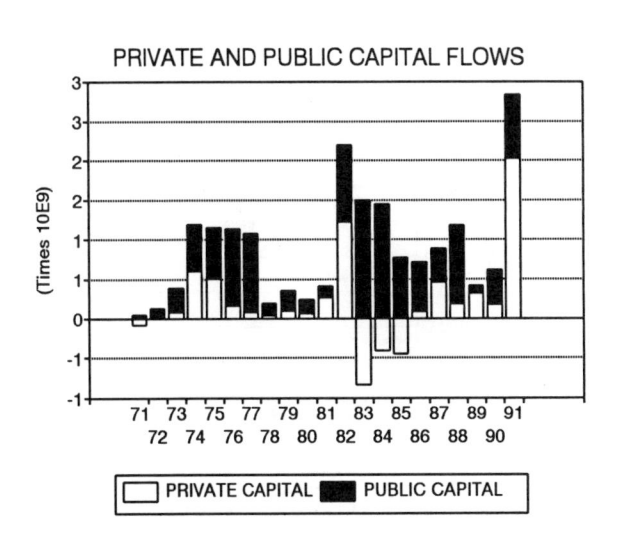

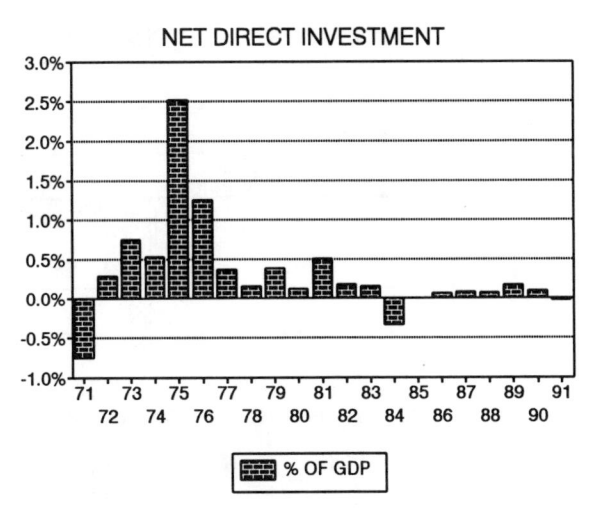

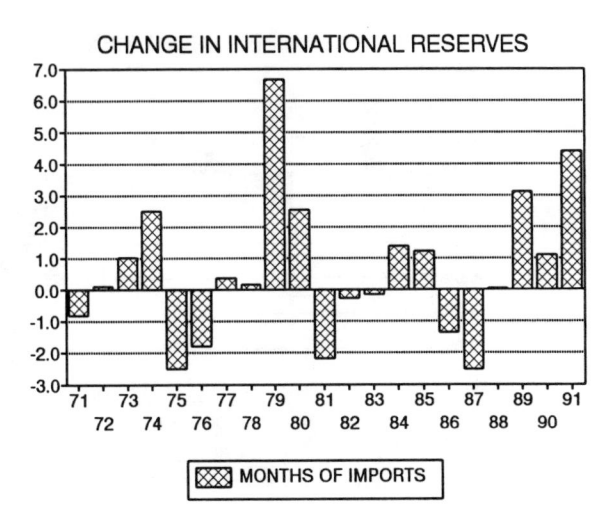

# PERU
## Balance of Payments in US Dollars

Statistics and Quantitative Analysis/IDB

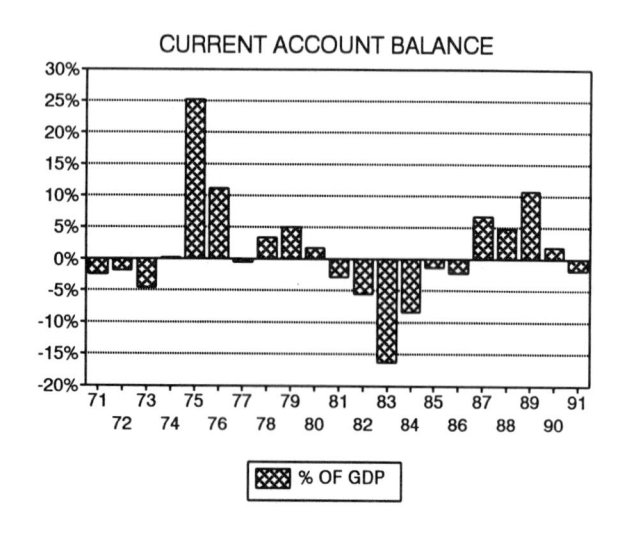

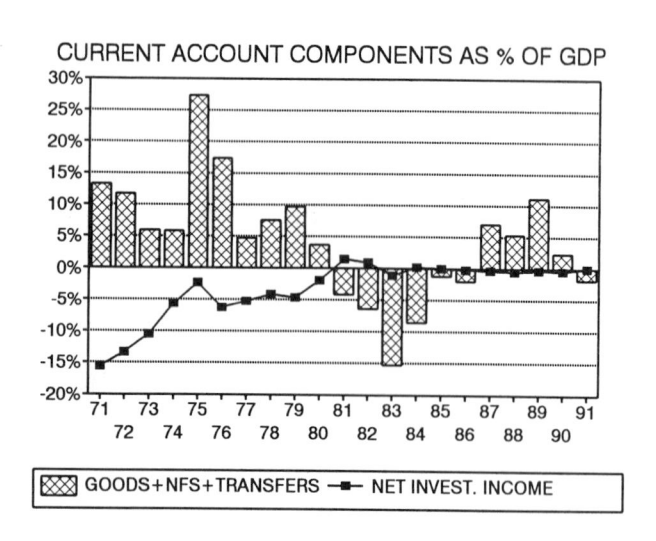

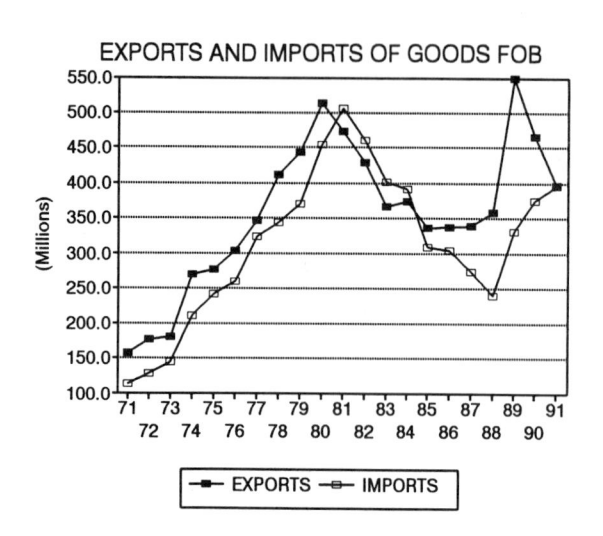

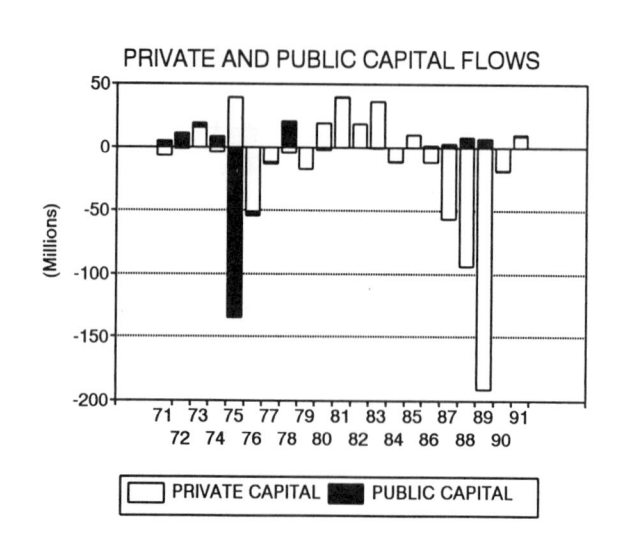

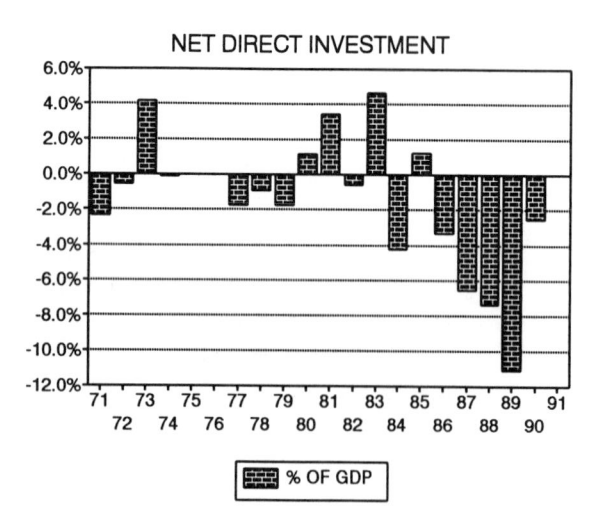

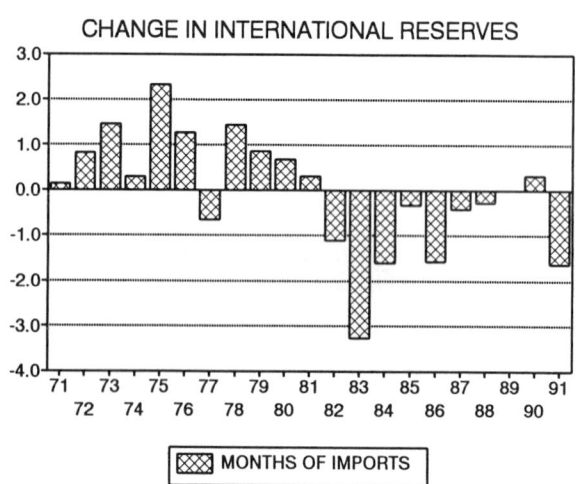

# SURINAME
## Balance of Payments in US Dollars

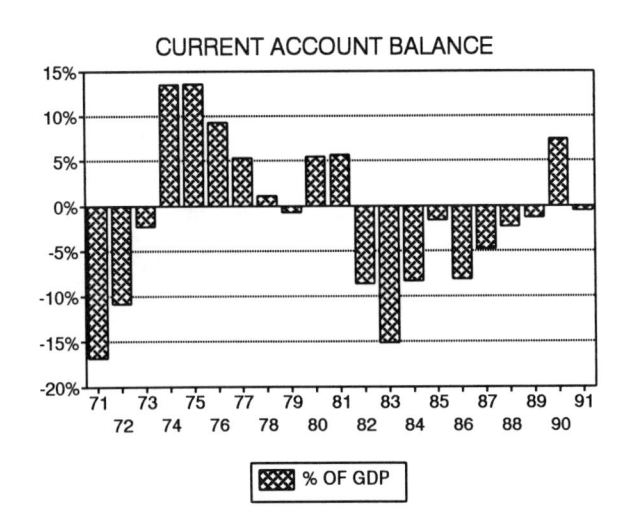

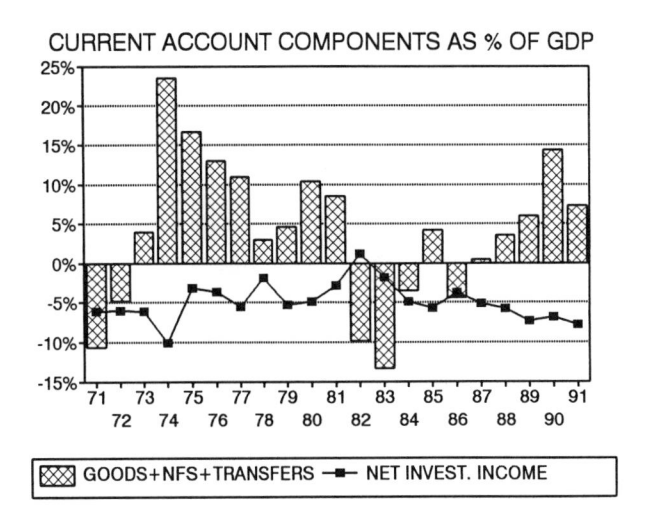

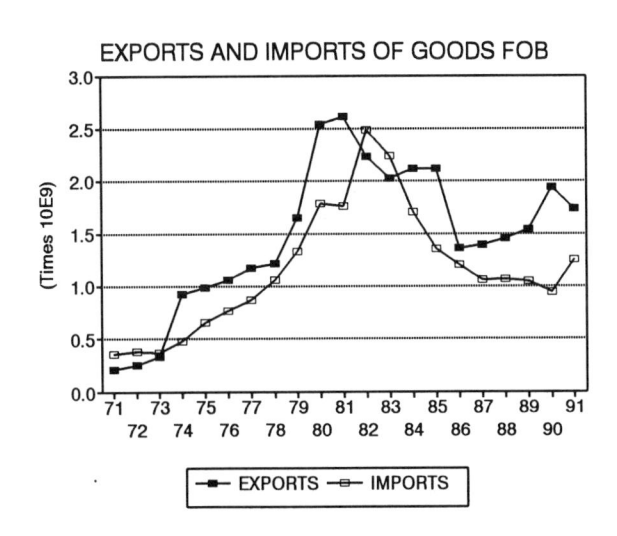

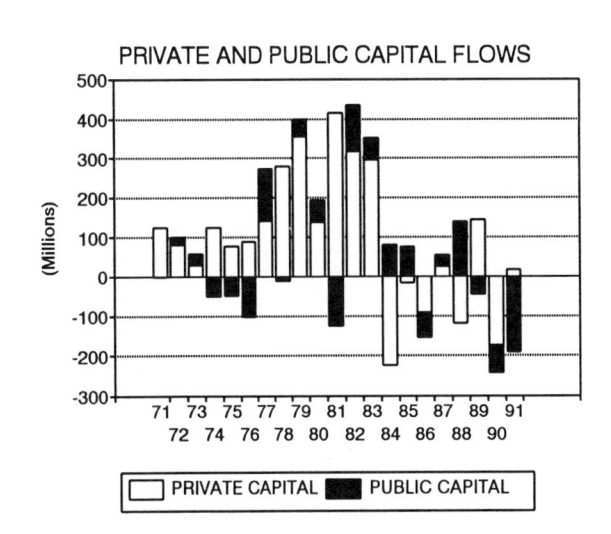

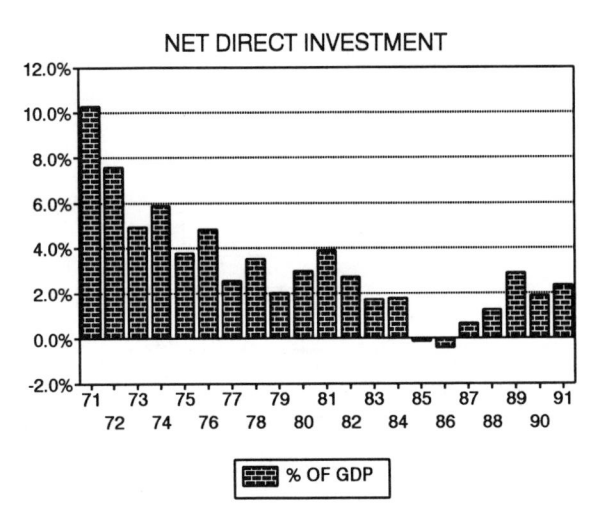

# TRINIDAD AND TOBAGO
## Balance of Payments in US Dollars

Statistics and Quantitative Analysis/IDB

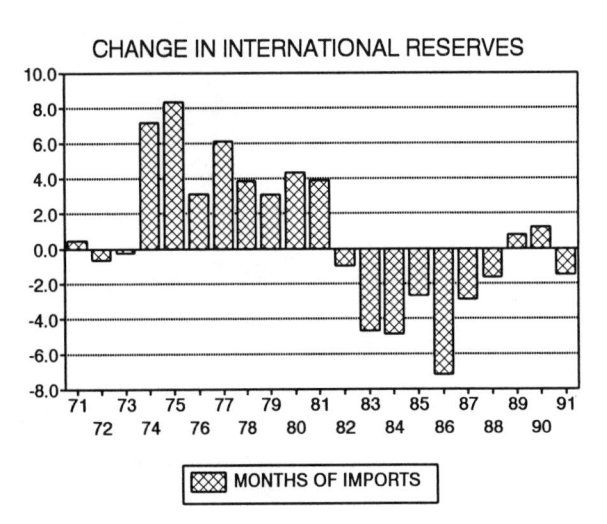

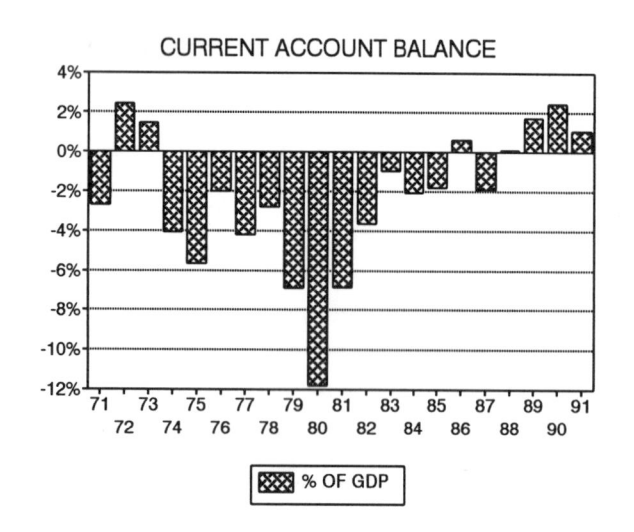

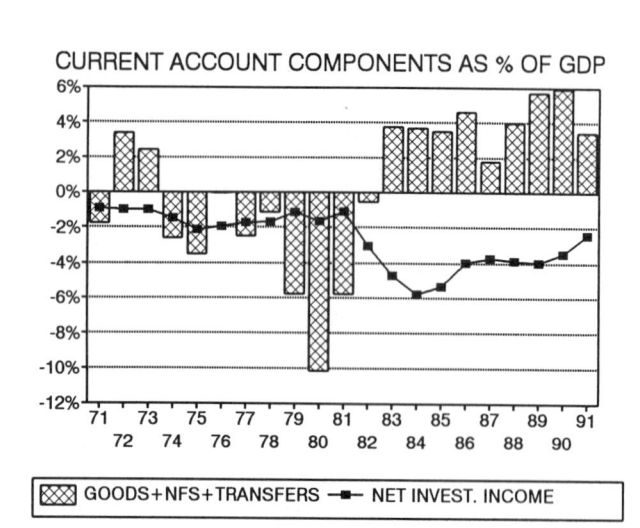

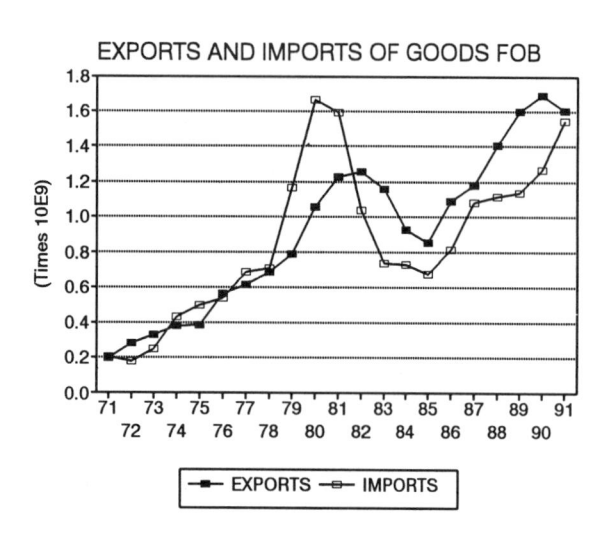

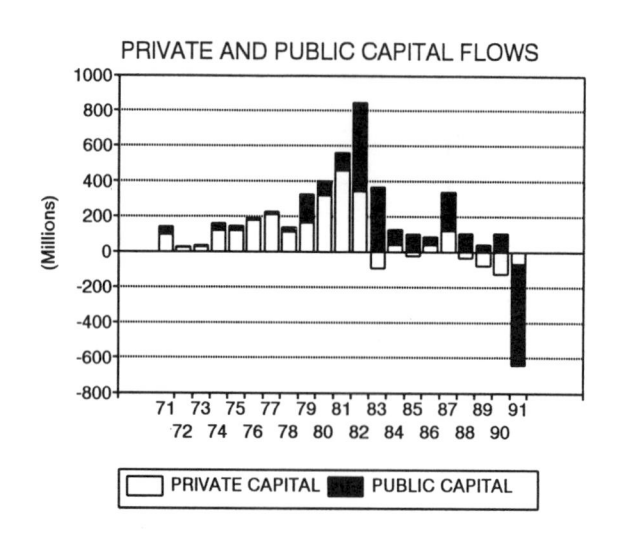

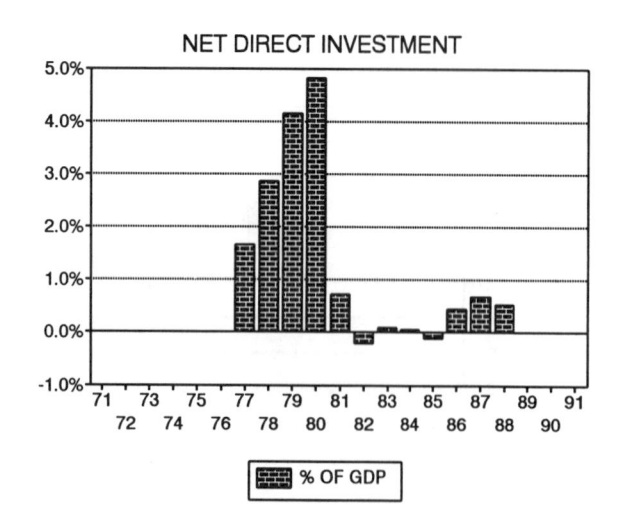

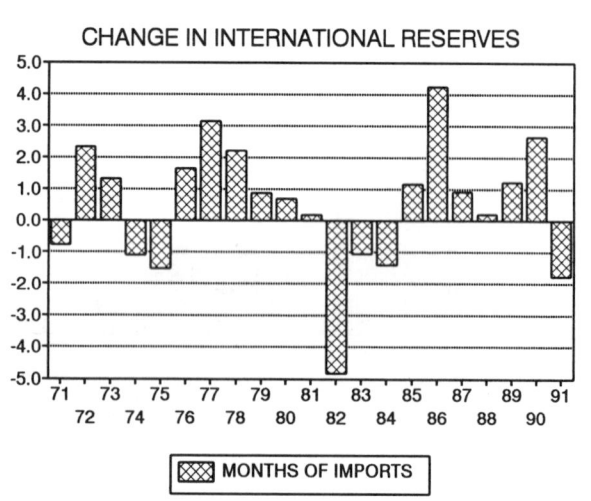

# URUGUAY
## Balance of Payments in US Dollars

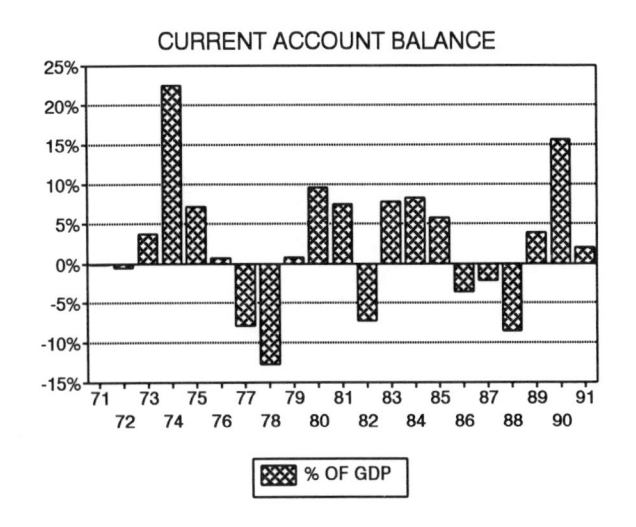

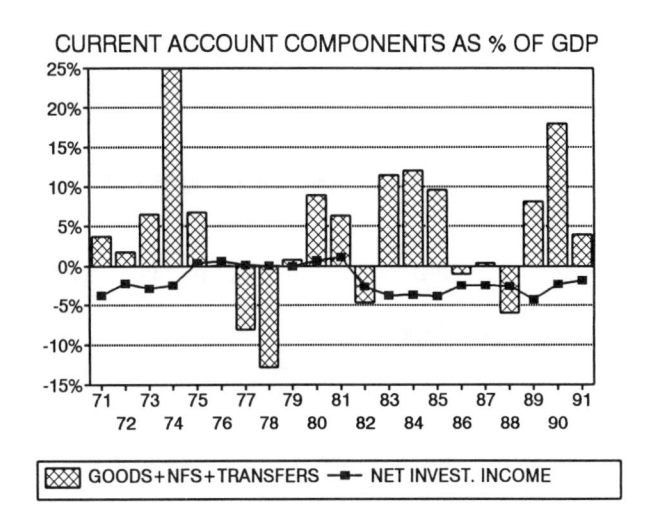

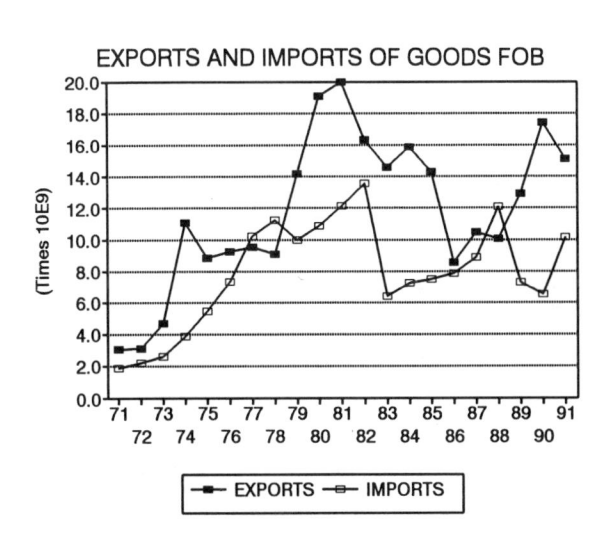

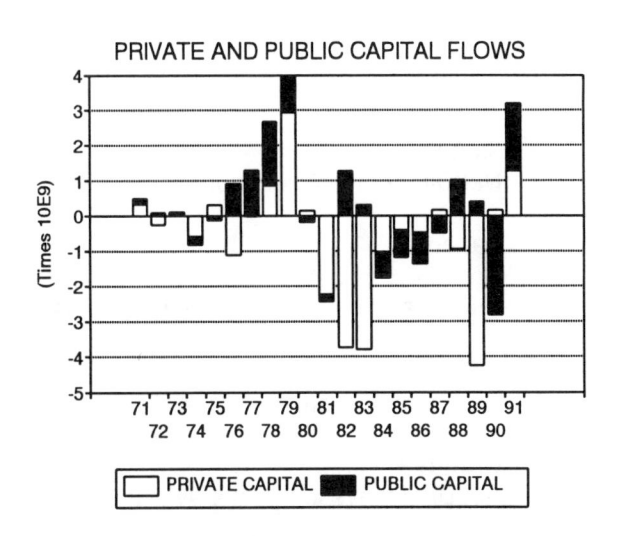

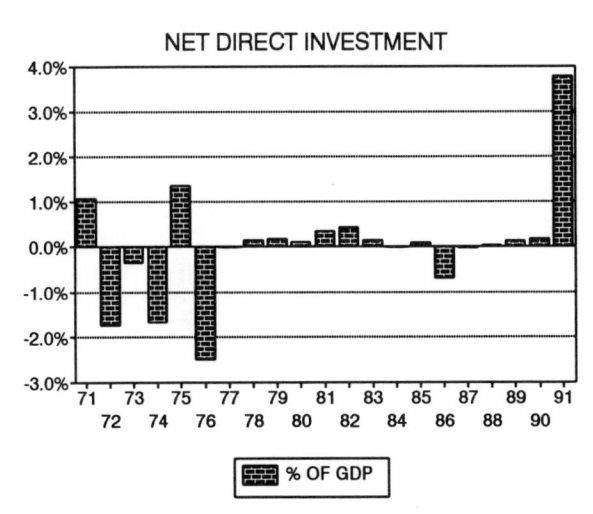

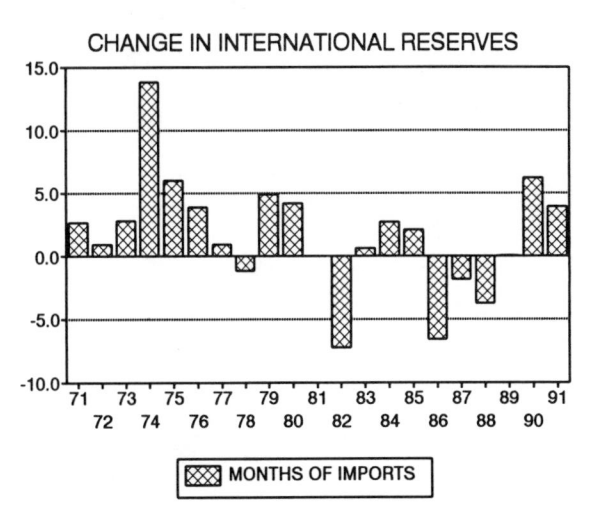

# VENEZUELA
## Balance of Payments in US Dollars

Statistics and Quantitative Analysis/IDB

## Country Notes on Balance of Payments

### Bahamas
Data are not available for 1971-1972.

### Barbados
Net Direct Investment not available for 1991.

### Brazil
Private and Public capital flows not available for 1990-1991.

### Costa Rica
Net Direct Investment not available for 1991.

### El Salvador
Net Direct Investment not available for 1990-1991.

### Guatemala
Net Direct Investment not available for 1990-1991.

### Guyana
Net Direct Investment not available since 1986.

### Haiti
Net Direct Investment not available for 1991.

### Jamaica
Net Direct Investment not available for 1991.

### Panama
Net Direct Investment not available for 1991.
Private and Public capital flows not available for 1991.

### Suriname
Net Direct Investment not available for 1991.

### Uruguay
Net Direct Investment not available for 1991.

# EXTERNAL TRADE

# EXTERNAL TRADE

The Export and Import structures for 1970 are based on US Dollar data for 1970 or the most proximate year in the 1970's. Similarly, the structures for 1990 are based on the most recent data available (see country notes at the end of this Section for more details).

## STRUCTURE OF EXPORTS OF GOODS

Data on Exports of Goods are from the United Nations Statistical Division (UNSTAT) Commodity Trade (COMTRADE) data base. The classification is based on aggregates of the Standard International Trade Classification (SITC), Revised Sections as follows:

1. *All Food and Agricultural Raw Materials*: SITC 0+1+2+4-(27+28) includes food, live animals, beverages and tobacco, oil and fats and primary raw commodities like wood, cork, silk, cotton etc..

2. *Fuels*: SITC 3 includes coal, crude and derivative oil, gas and electric energy.

3. *Ores and Metals*: SITC (27+28)+(67+68) includes ferrous and non-ferrous metal scraps, tubes pipes etc., copper, zinc etc..

4. *Manufactured Goods*: SITC 5+6+7+8-(67+68) includes chemicals, machinery and transport equipment, and other manufactured goods.

SITC, Revised includes in *All Food and Agricultural Raw Materials*, products which may be processed to some degree. For example, SITC 0 (Food and Live Animals) makes no distinction between green tea and tea which has been processed; likewise, SITC 1 (Beverages and Tobacco) includes tobacco manufactures and alcoholic beverages. This is unlike the International Standard Industrial Classification (ISIC) which includes food, beverage and tobacco processing in the manufacturing sector.

In the graphs for export composition, ALL FOOD = *All Food and Agricultural Raw Materials*. The structure is calculated as a percent of the sum of the four categories.

## STRUCTURE OF IMPORTS OF GOODS

Import data have been collected directly from the countries or provided by the Economic Comission for Latin America and the Caribbean (ECLAC). The classifications used are those developed by the countries themselves or by ECLAC based on the United Nations Classification by Broad Economic Categories (BEC). The Imports of Goods have been classified in the four following categories:

1. *Consumption Goods*

2. *Capital Goods*

3. *Oil*, including crude petroleum as well as derivative products (where available)

4. *Non-oil Intermediate Goods*

In the graphs for import composition, CON = *Consumption Goods*; CAP = *Capital Goods*; OIL = *Oil* and INT = *Non-oil Intermediate Goods*. In cases where Oil is not available separately, Intermediate Goods (INT) includes Oil. The structure is calculated as a percent of the sum of the four categories.

# LATIN AMERICA
## External Trade in US Dollars

### EXPORT STRUCTURE : 1970

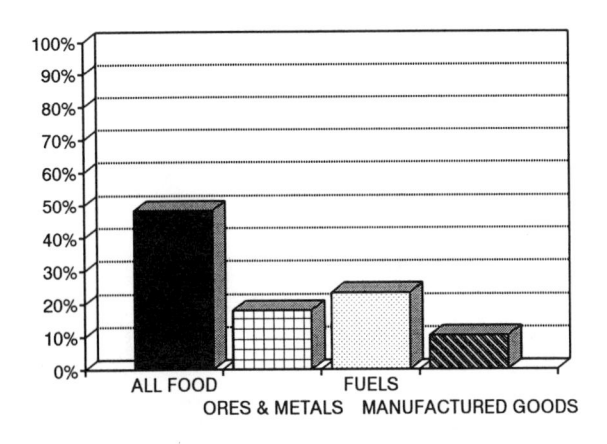

### IMPORT STRUCTURE : 1970

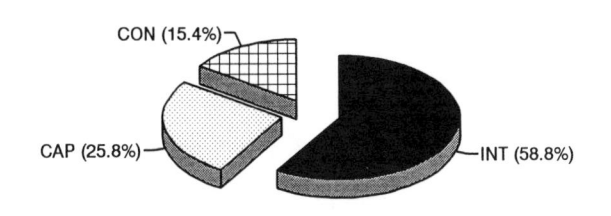

- CON (15.4%)
- CAP (25.8%)
- INT (58.8%)

### 1980

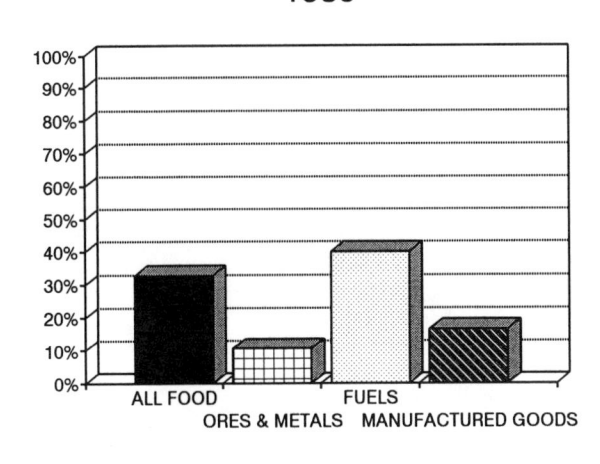

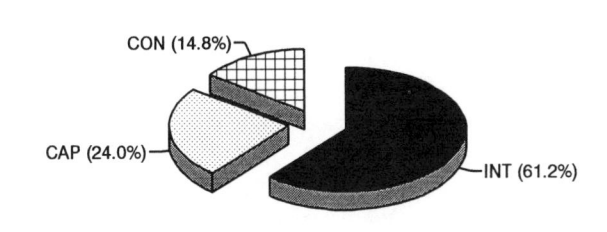

- CON (14.8%)
- CAP (24.0%)
- INT (61.2%)

### 1990

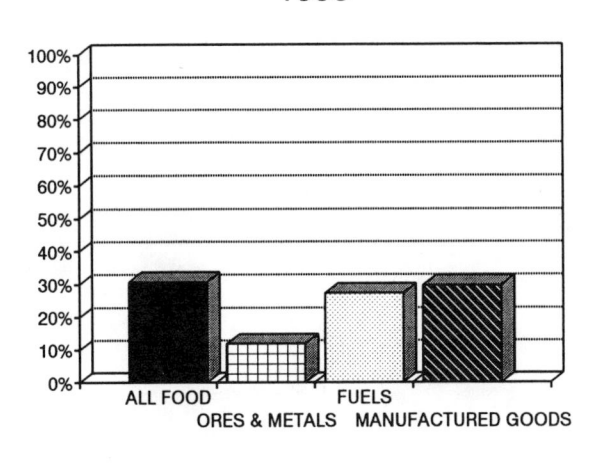

- CON (16.9%)
- CAP (22.8%)
- INT (60.3%)

Statistics and Quantitative Analysis/IDB

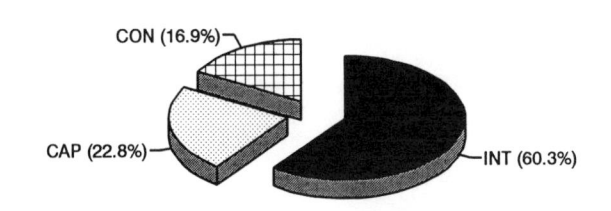

# ARGENTINA
## External Trade in US Dollars

### EXPORT STRUCTURE : 1970

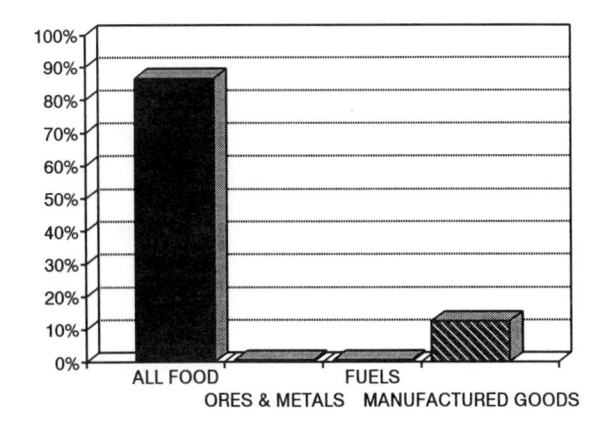

### IMPORT STRUCTURE : 1970

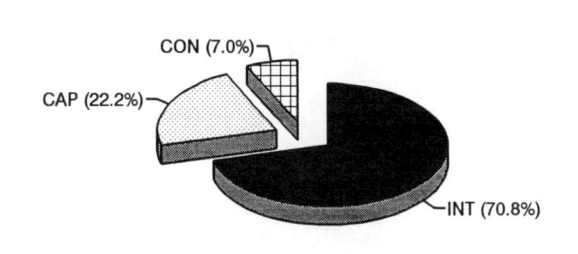

- CON (7.0%)
- CAP (22.2%)
- INT (70.8%)

### 1980

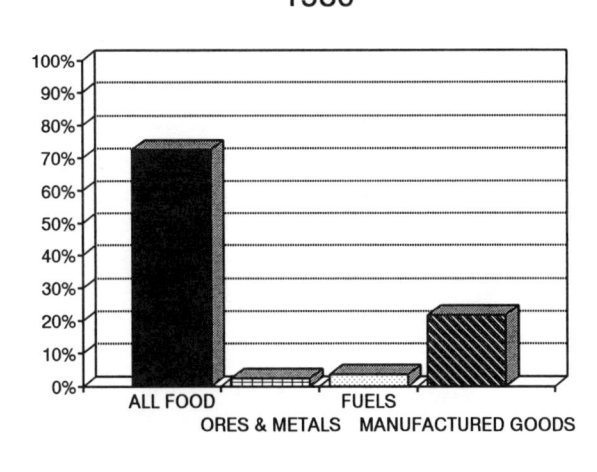

### 1980

- CON (19.2%)
- CAP (28.0%)
- INT (52.8%)

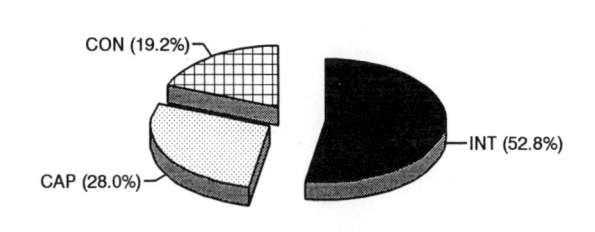

### 1990

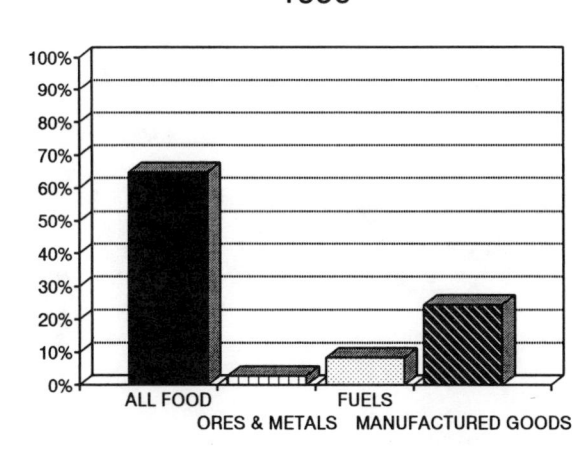

### 1990

- CON (8.0%)
- CAP (19.8%)
- INT (72.1%)

Statistics and Quantitative Analysis/IDB

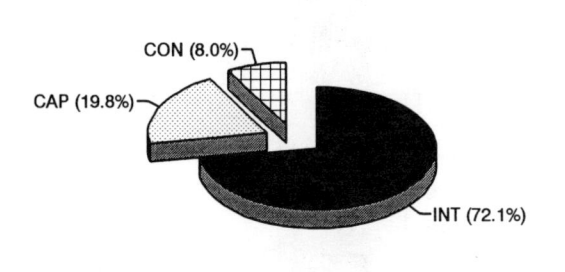

# BAHAMAS
## External Trade in US Dollars

### EXPORT STRUCTURE : 1970

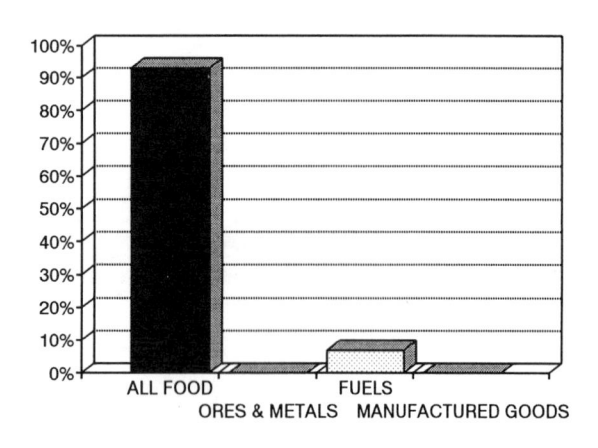

### 1980

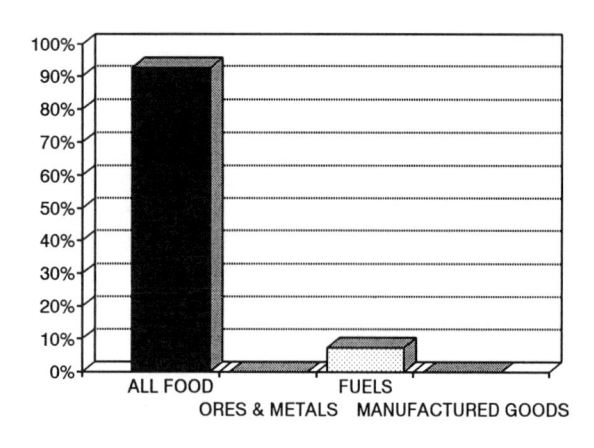

### 1990

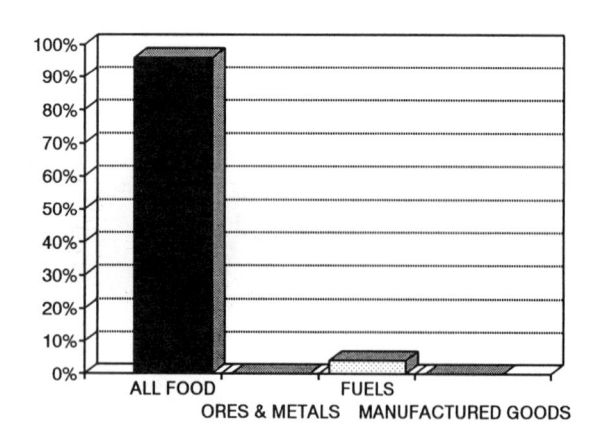

# BARBADOS
## External Trade in US Dollars

### EXPORT STRUCTURE : 1970

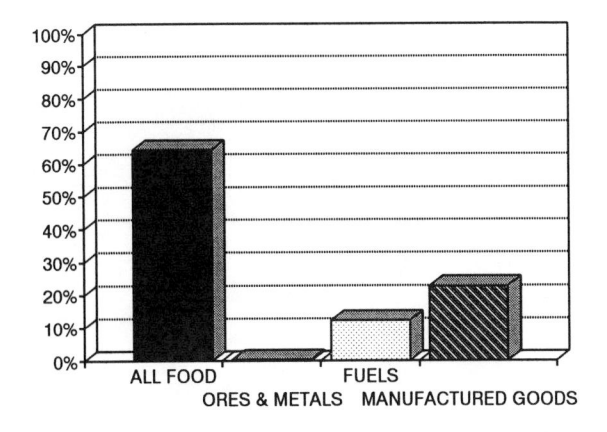

### IMPORT STRUCTURE : 1970

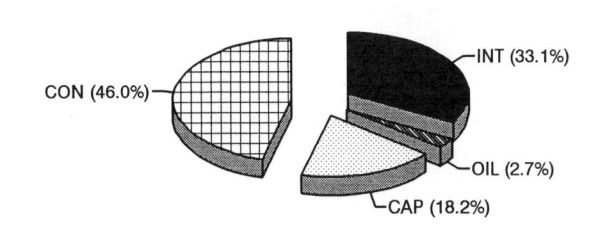

CON (46.0%)
INT (33.1%)
OIL (2.7%)
CAP (18.2%)

### 1980

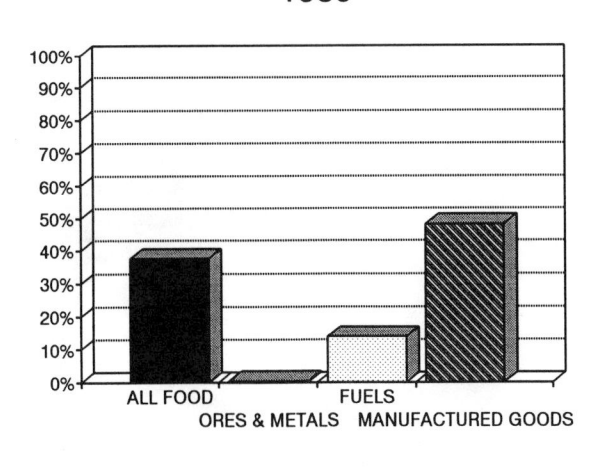

### 1980

CON (36.7%)
INT (34.6%)
OIL (11.6%)
CAP (17.2%)

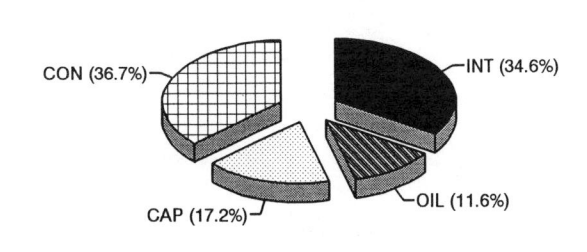

### 1990

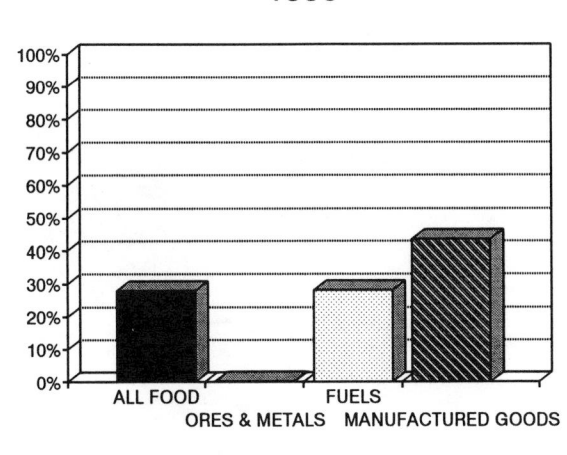

### 1990

CON (38.4%)
INT (32.4%)
OIL (7.1%)
CAP (22.1%)

Statistics and Quantitative Analysis/IDB

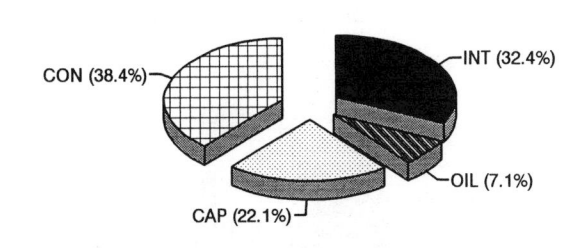

# BOLIVIA
## External Trade in US Dollars

### EXPORT STRUCTURE : 1970

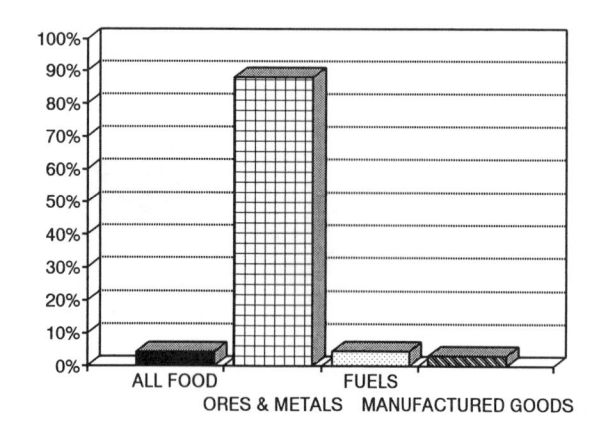

### IMPORT STRUCTURE : 1970

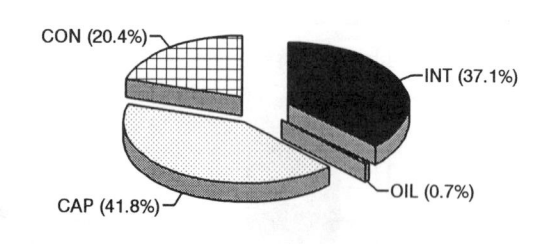

CON (20.4%)
INT (37.1%)
OIL (0.7%)
CAP (41.8%)

### 1980

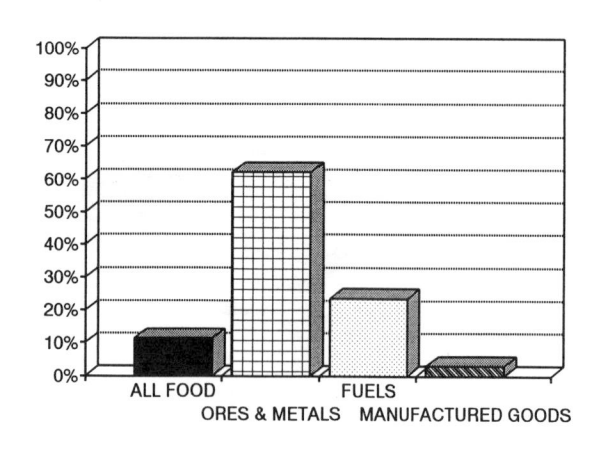

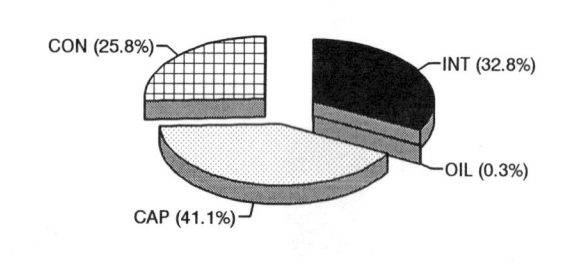

CON (25.8%)
INT (32.8%)
OIL (0.3%)
CAP (41.1%)

### 1990

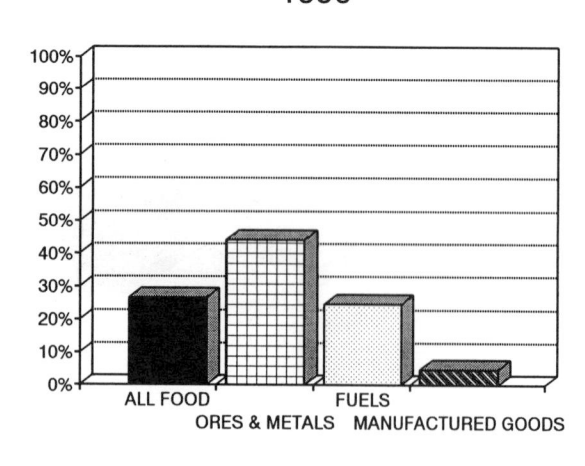

CON (20.8%)
INT (38.1%)
OIL (0.5%)
CAP (40.6%)

Statistics and Quantitative Analysis/IDB

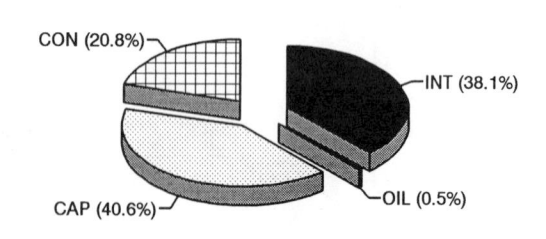

# BRAZIL
## External Trade in US Dollars

### EXPORT STRUCTURE : 1970

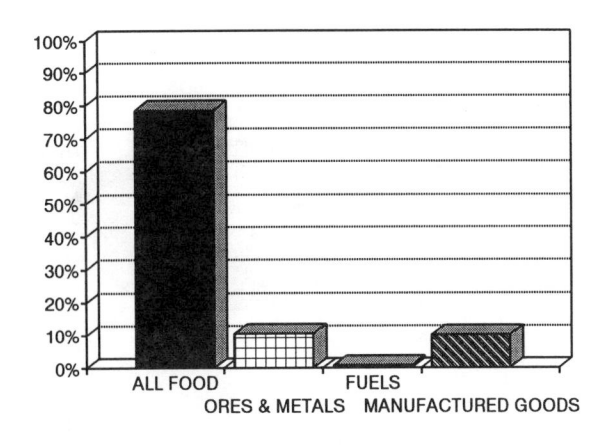

### IMPORT STRUCTURE : 1970

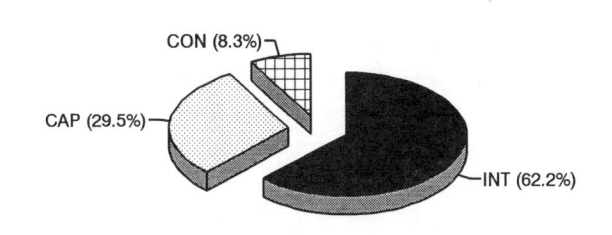

- CON (8.3%)
- CAP (29.5%)
- INT (62.2%)

### 1980

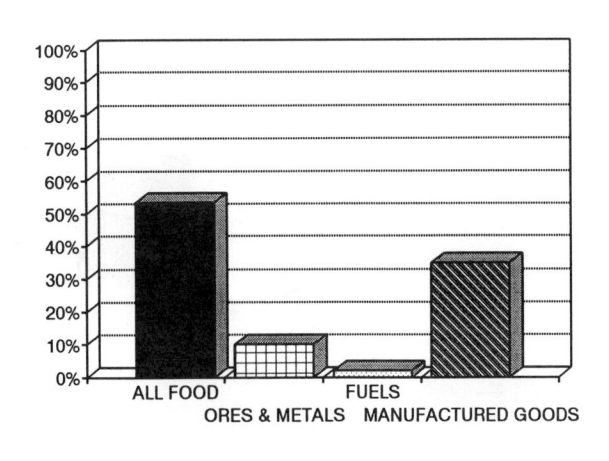

### 1980

- CON (2.4%)
- CAP (15.4%)
- INT (82.3%)

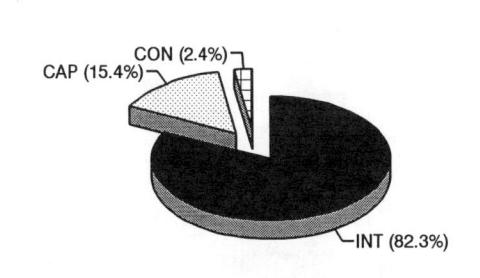

### 1990

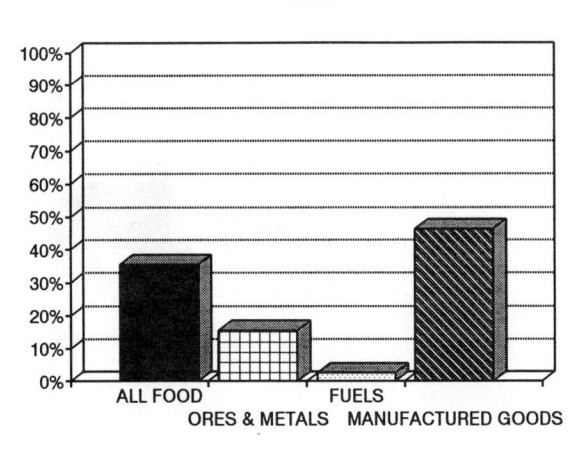

### 1990

- CON (10.4%)
- CAP (15.6%)
- INT (74.0%)

Statistics and Quantitative Analysis/IDB

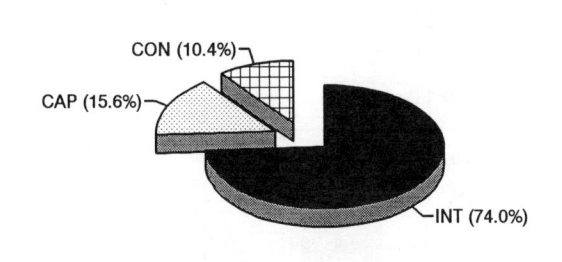

# CHILE
## External Trade in US Dollars

### EXPORT STRUCTURE : 1970

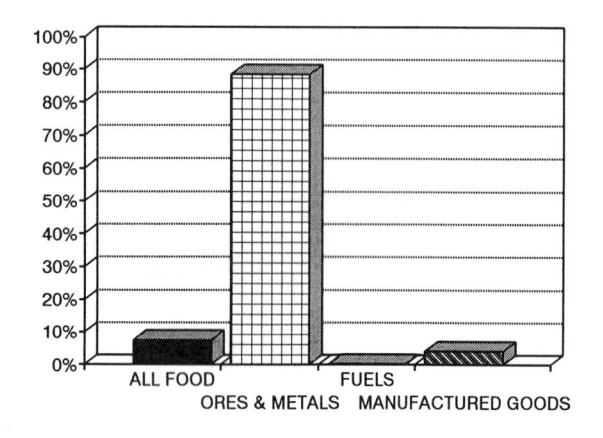

### IMPORT STRUCTURE : 1970

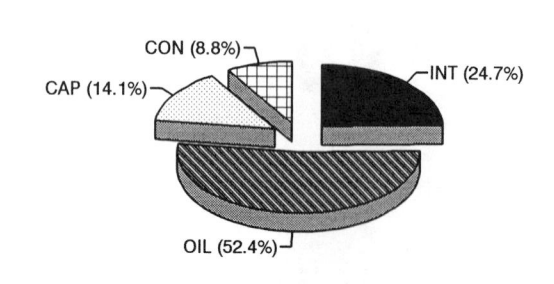

- CON (8.8%)
- CAP (14.1%)
- OIL (52.4%)
- INT (24.7%)

### 1980

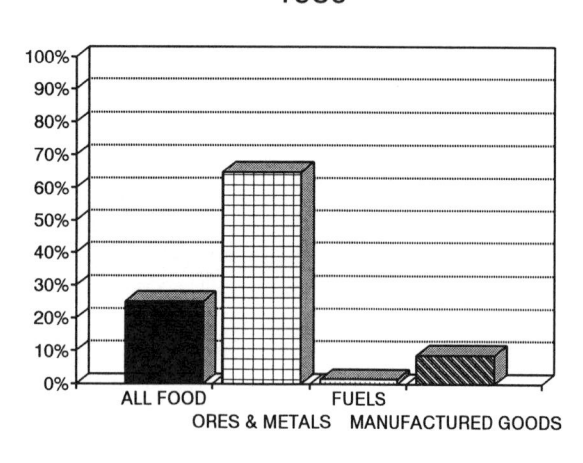

### 1980

- CON (33.7%)
- CAP (20.7%)
- INT (29.9%)
- OIL (15.7%)

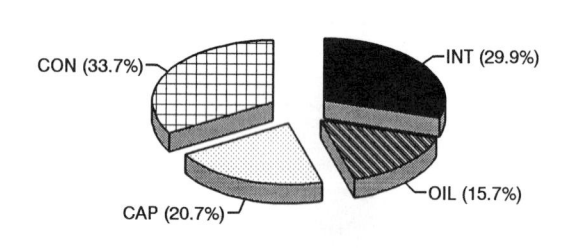

### 1990

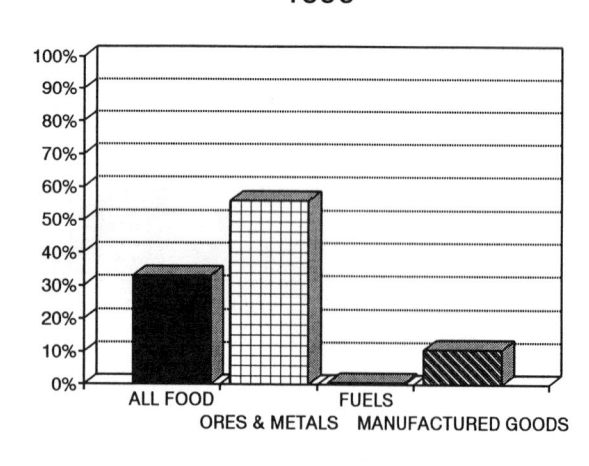

### 1990

- CON (20.9%)
- CAP (27.3%)
- INT (40.4%)
- OIL (11.4%)

Statistics and Quantitative Analysis/IDB

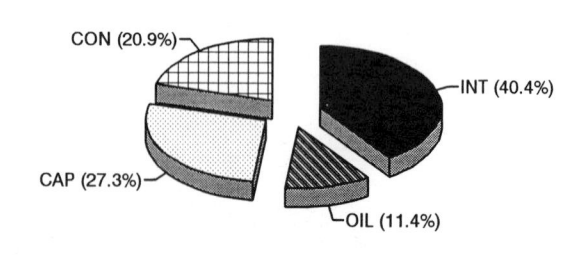

# COLOMBIA
### External Trade in US Dollars

## EXPORT STRUCTURE : 1970

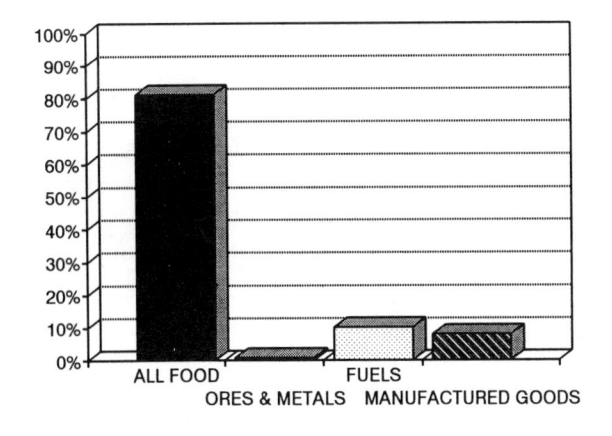

## IMPORT STRUCTURE : 1970

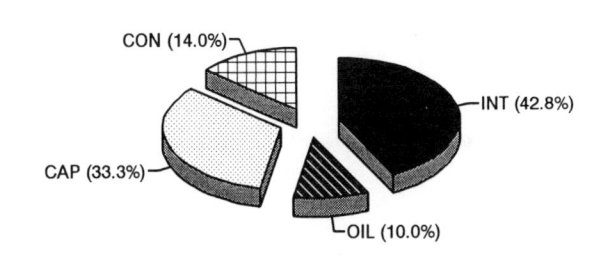

- CON (14.0%)
- CAP (33.3%)
- OIL (10.0%)
- INT (42.8%)

## 1980

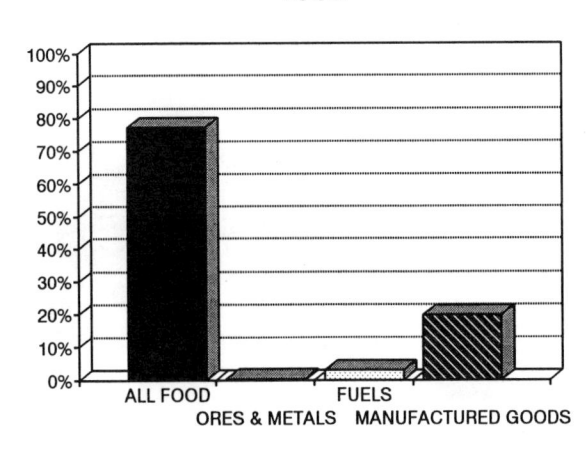

## 1980

- CON (13.3%)
- CAP (34.0%)
- OIL (12.1%)
- INT (40.7%)

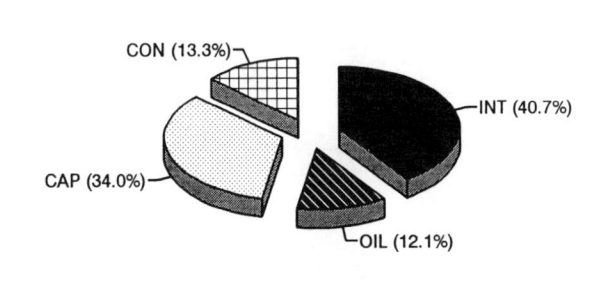

## 1990

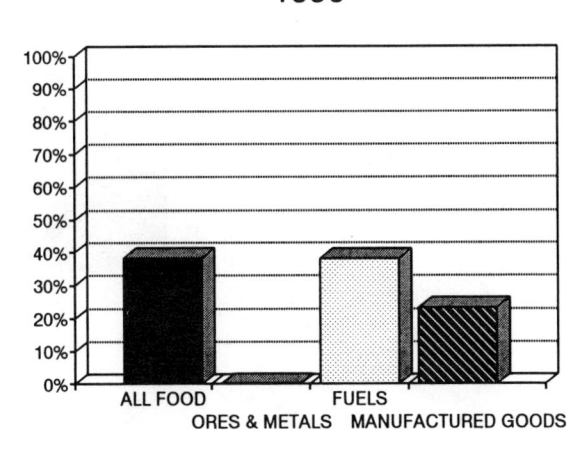

## 1990

- CON (10.0%)
- CAP (36.4%)
- OIL (5.7%)
- INT (47.9%)

Statistics and Quantitative Analysis/IDB

77

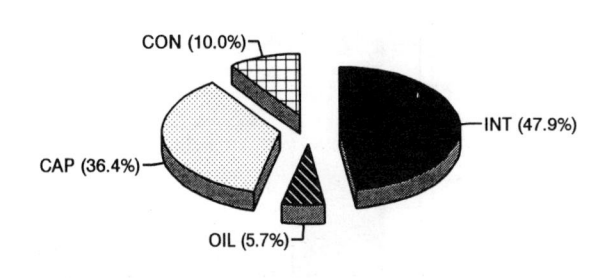

# COSTA RICA
## External Trade in US Dollars

### EXPORT STRUCTURE : 1970

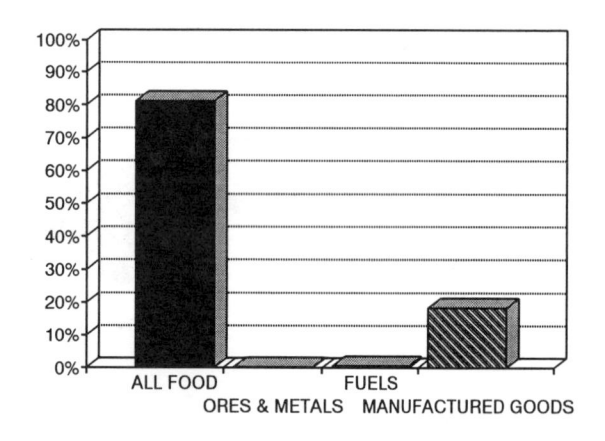

### IMPORT STRUCTURE : 1970

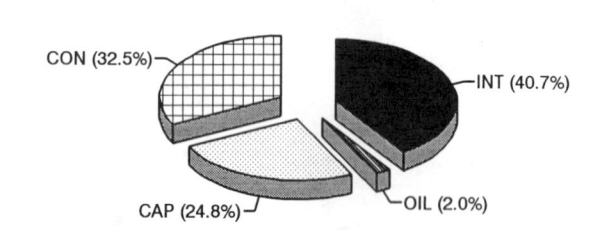

CON (32.5%)
INT (40.7%)
CAP (24.8%)
OIL (2.0%)

### 1980

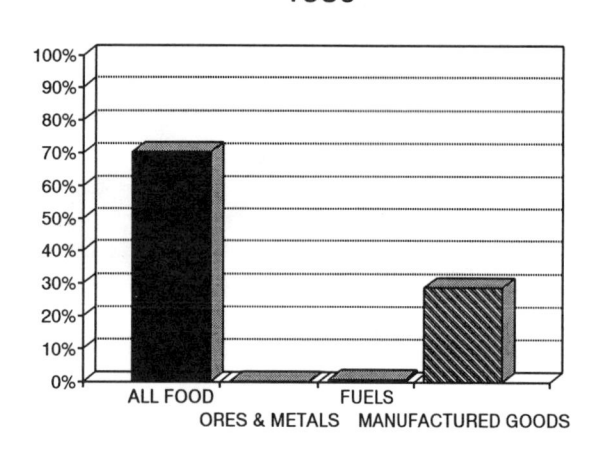

### 1980

CON (25.8%)
INT (38.5%)
CAP (21.6%)
OIL (14.1%)

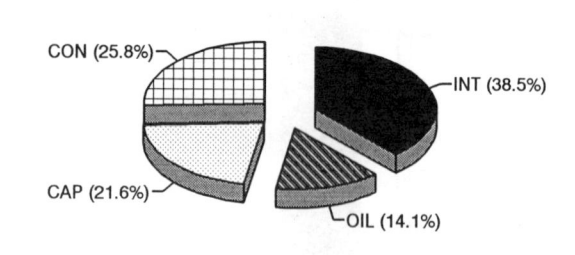

### 1990

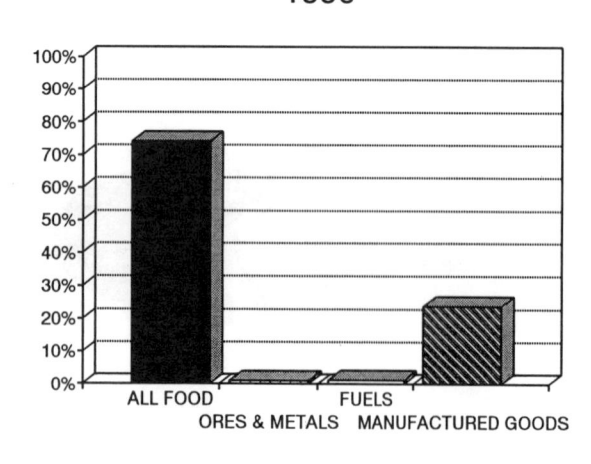

### 1990

CON (22.9%)
INT (46.7%)
CAP (20.8%)
OIL (9.6%)

Statistics and Quantitative Analysis/IDB

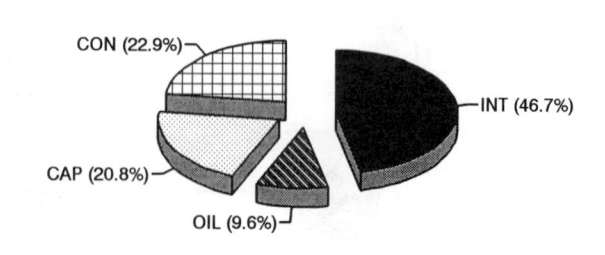

# DOMINICAN REPUBLIC
## External Trade in US Dollars

### EXPORT STRUCTURE : 1970

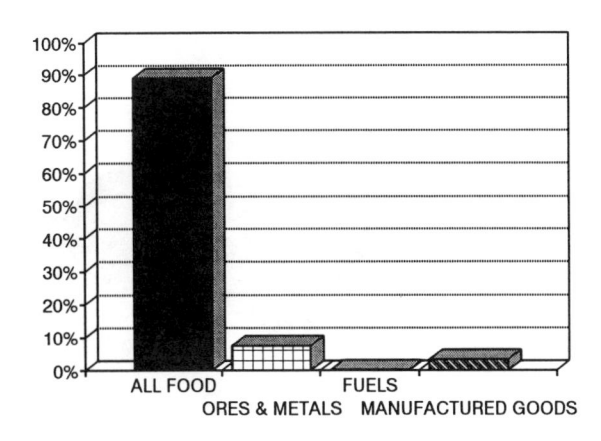

### IMPORT STRUCTURE : 1970
- CON (12.7%)
- CAP (23.0%)
- INT (64.4%)

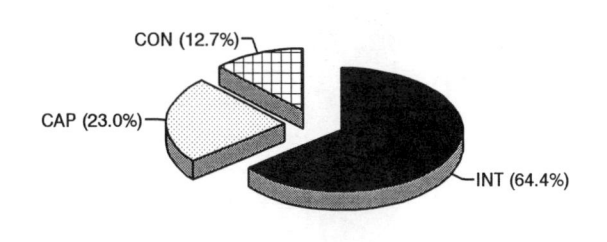

### 1980

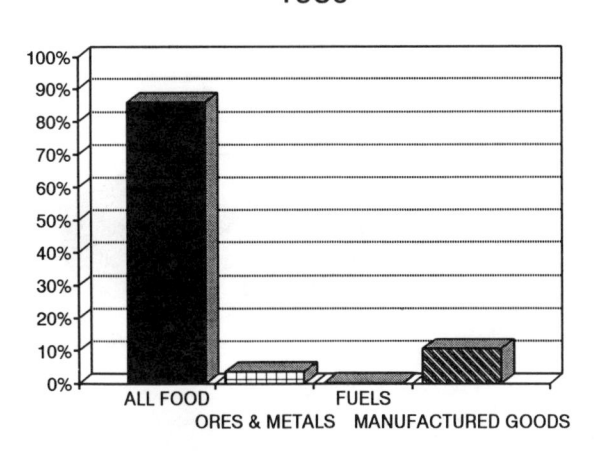

### 1980
- CON (13.0%)
- CAP (17.4%)
- INT (69.6%)

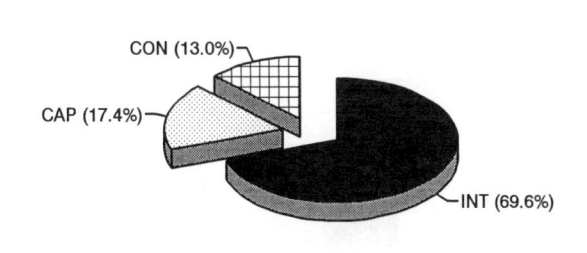

### 1990

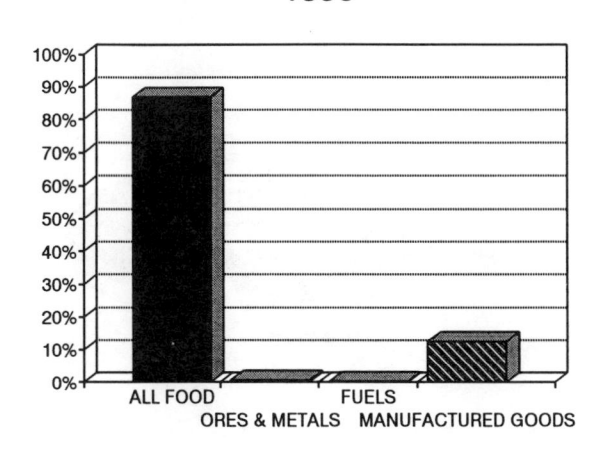

### 1990
- INT (9.8%)
- CAP (16.8%)
- CON (73.4%)

Statistics and Quantitative Analysis/IDB

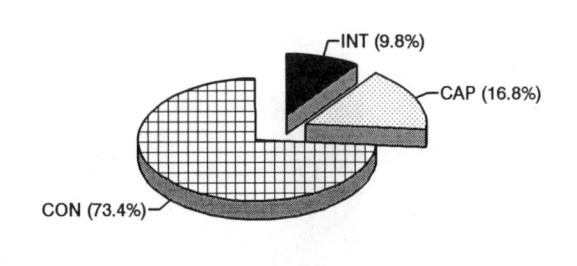

# ECUADOR
**External Trade in US Dollars**

## EXPORT STRUCTURE : 1970

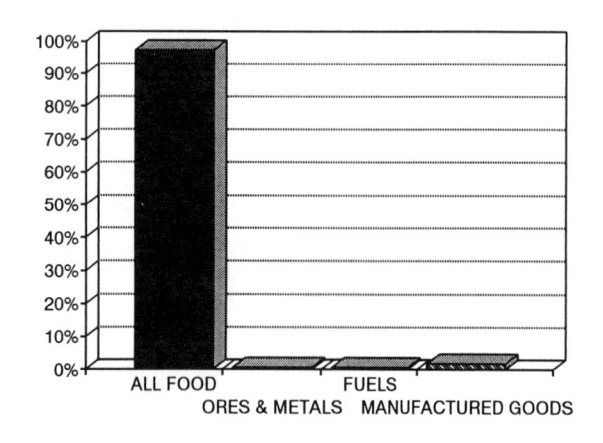

## IMPORT STRUCTURE : 1970

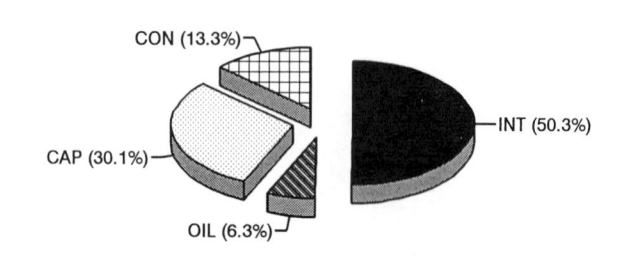

- CON (13.3%)
- CAP (30.1%)
- OIL (6.3%)
- INT (50.3%)

## 1980

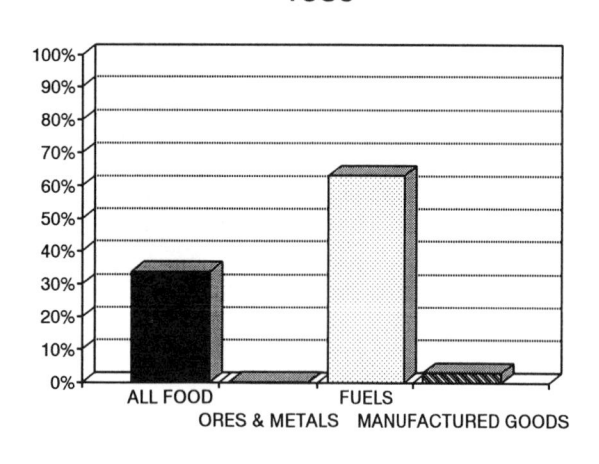

- CON (8.5%)
- CAP (40.7%)
- OIL (9.8%)
- INT (41.0%)

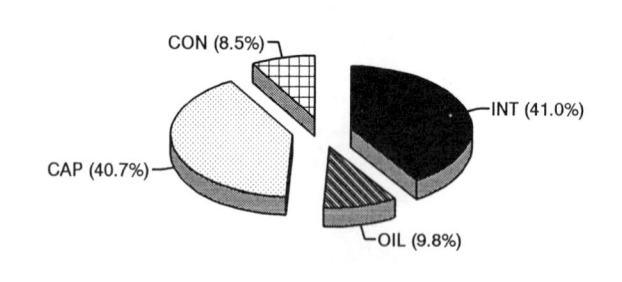

## 1990

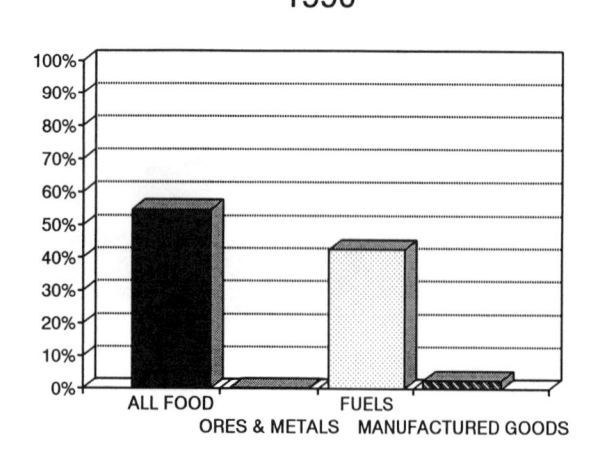

- CON (9.6%)
- CAP (32.6%)
- OIL (4.9%)
- INT (52.8%)

Statistics and Quantitative Analysis/IDB

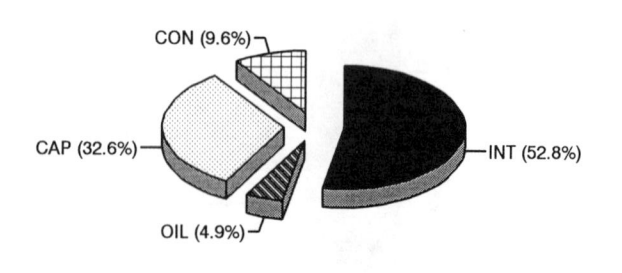

# EL SALVADOR
## External Trade in US Dollars

### EXPORT STRUCTURE : 1970

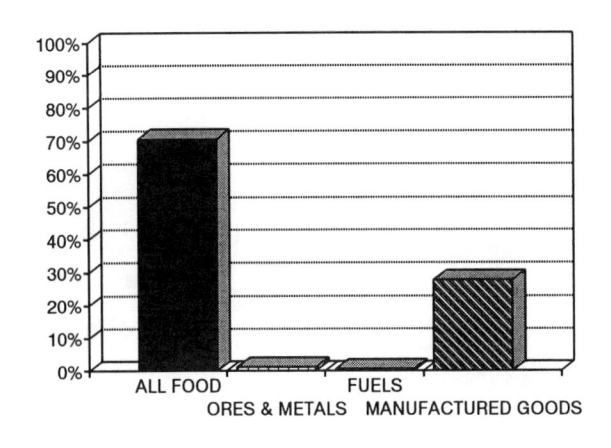

### IMPORT STRUCTURE : 1970

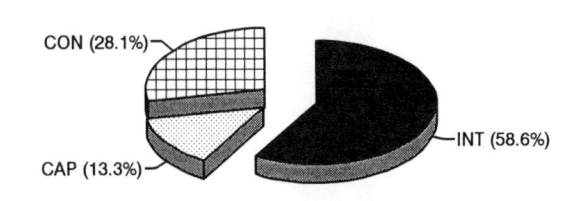

CON (28.1%)
CAP (13.3%)
INT (58.6%)

### 1980

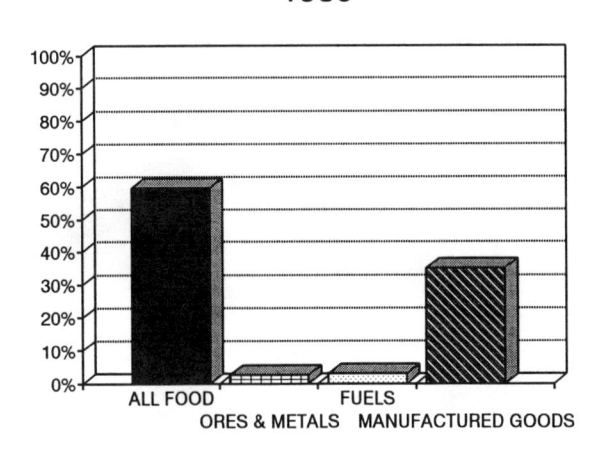

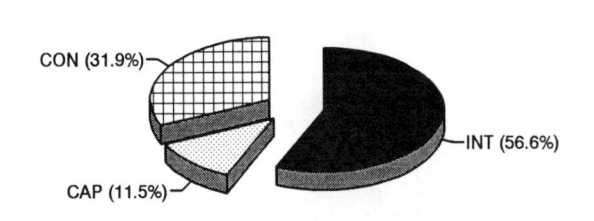

CON (31.9%)
CAP (11.5%)
INT (56.6%)

### 1990

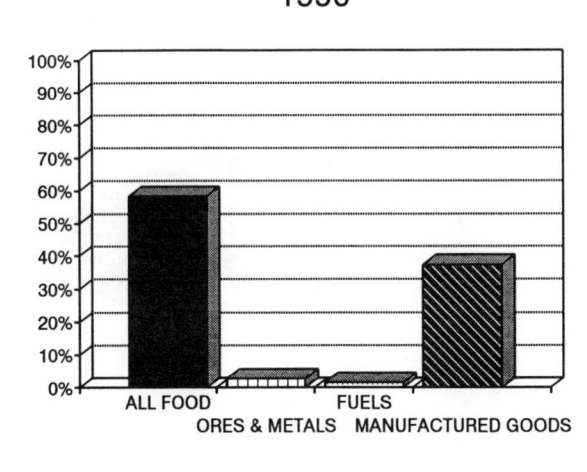

CON (31.6%)
CAP (18.6%)
INT (49.8%)

Statistics and Quantitative Analysis/IDB

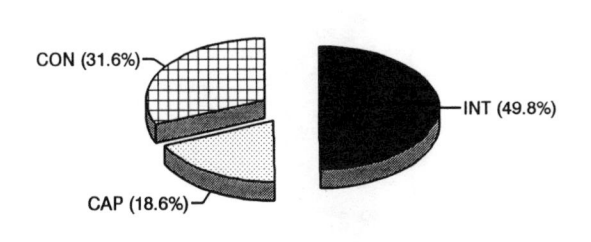

# GUATEMALA
## External Trade in US Dollars

### EXPORT STRUCTURE : 1970

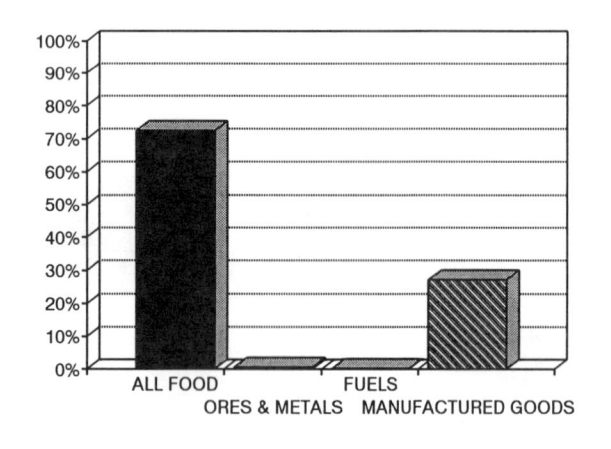

### IMPORT STRUCTURE : 1970

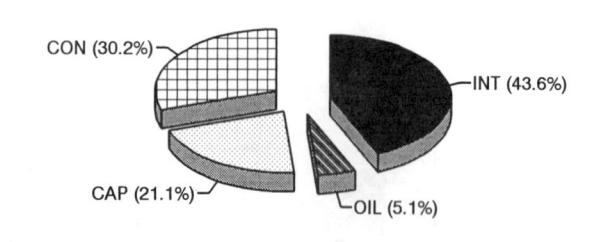

CON (30.2%)
CAP (21.1%)
INT (43.6%)
OIL (5.1%)

### 1980

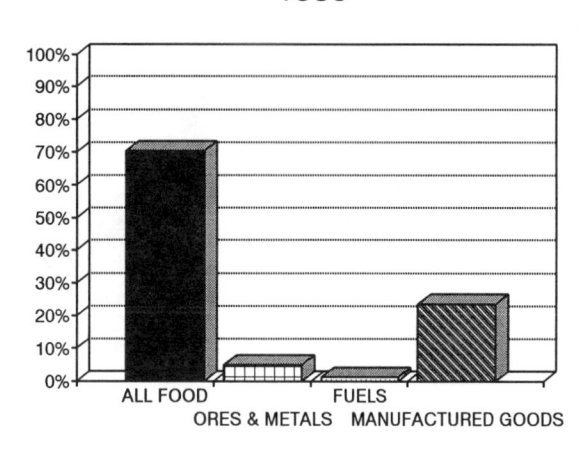

### 1980

CON (21.6%)
CAP (18.1%)
INT (39.0%)
OIL (21.4%)

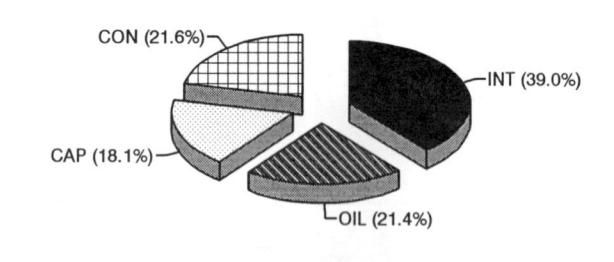

### 1990

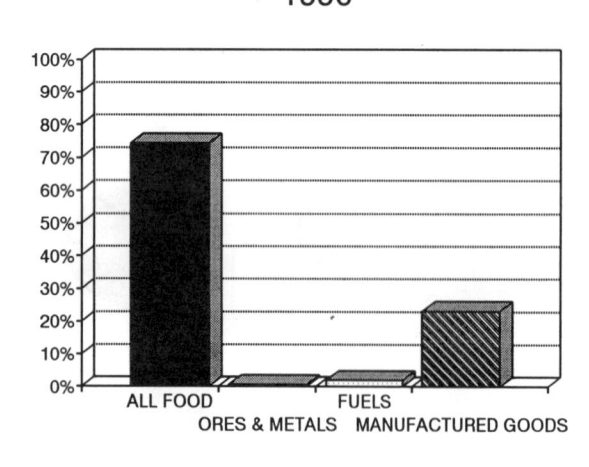

### 1990

CON (19.0%)
CAP (21.4%)
OIL (7.2%)
INT (52.4%)

Statistics and Quantitative Analysis/IDB

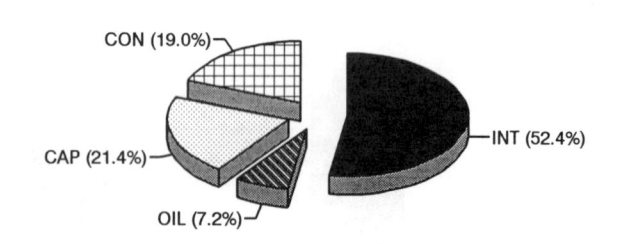

# GUYANA
## External Trade in US Dollars

### EXPORT STRUCTURE : 1970

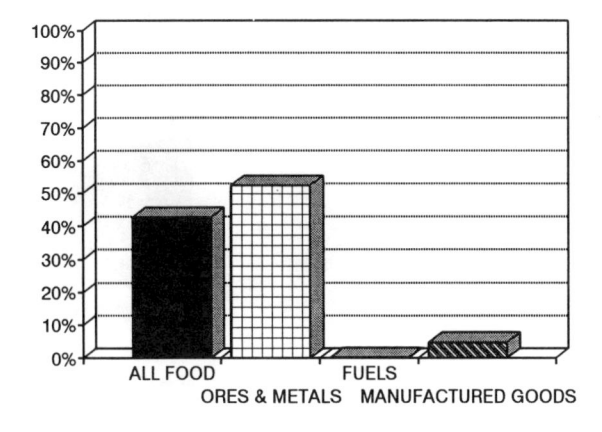

### IMPORT STRUCTURE : 1970

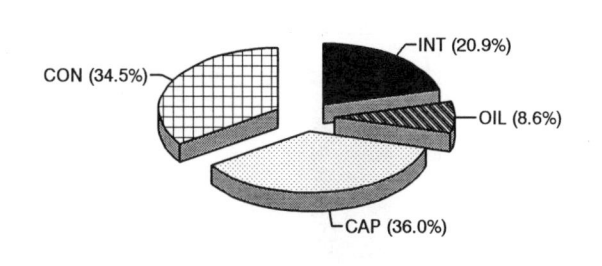

- CON (34.5%)
- INT (20.9%)
- OIL (8.6%)
- CAP (36.0%)

### 1980

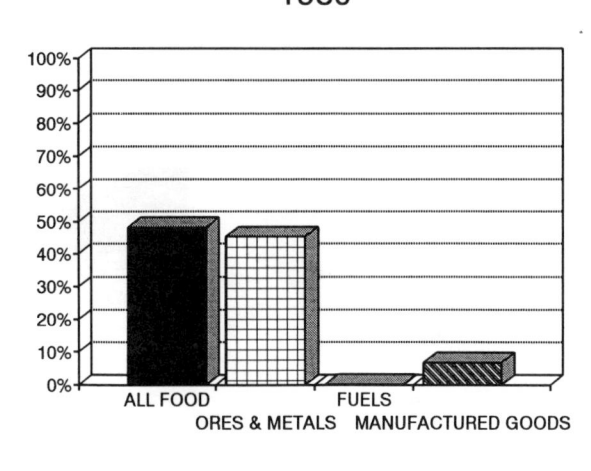

### 1980

- CON (12.9%)
- CAP (19.2%)
- INT (32.0%)
- OIL (35.9%)

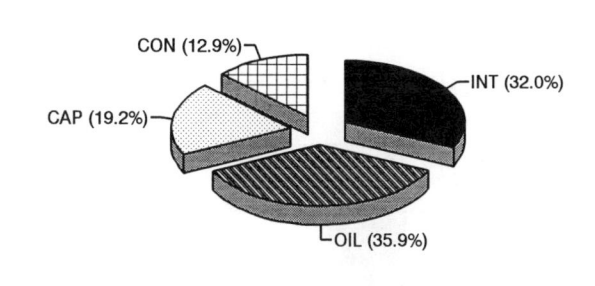

### 1990

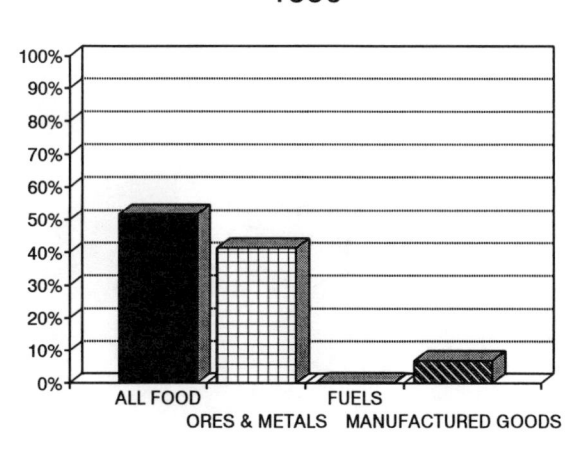

### 1990

- CON (12.1%)
- CAP (45.6%)
- INT (19.3%)
- OIL (23.0%)

Statistics and Quantitative Analysis/IDB

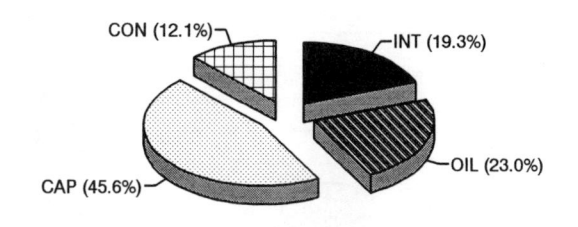

# HAITI
## External Trade in US Dollars

### EXPORT STRUCTURE : 1970

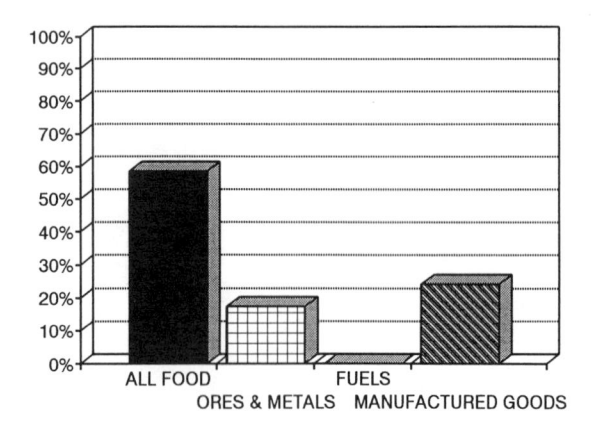

### IMPORT STRUCTURE : 1970

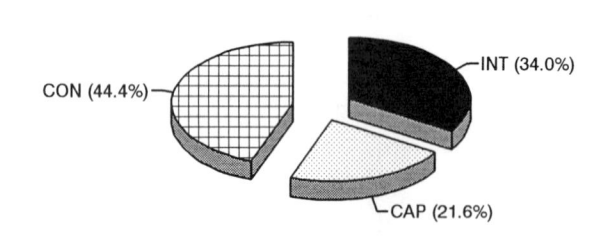

CON (44.4%)
INT (34.0%)
CAP (21.6%)

### 1980

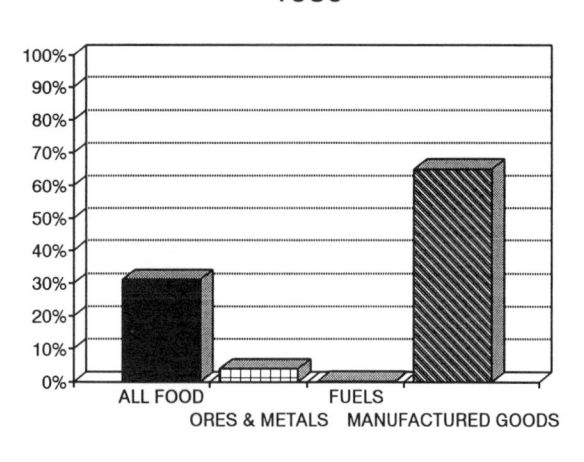

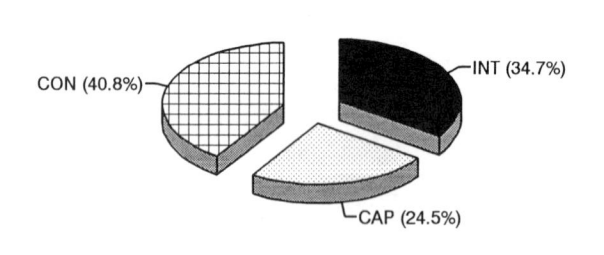

CON (40.8%)
INT (34.7%)
CAP (24.5%)

### 1990

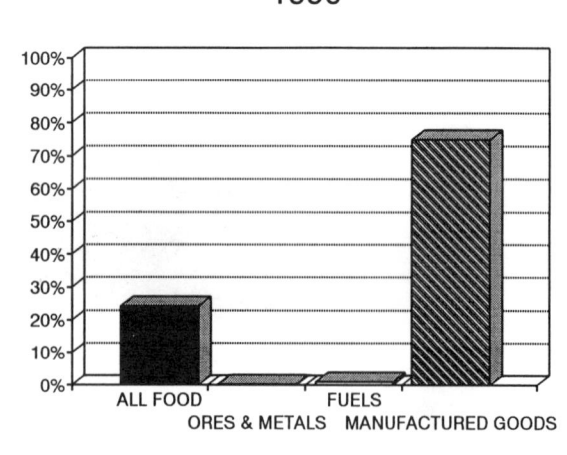

CON (28.8%)
CAP (11.3%)
INT (59.9%)

Statistics and Quantitative Analysis/IDB

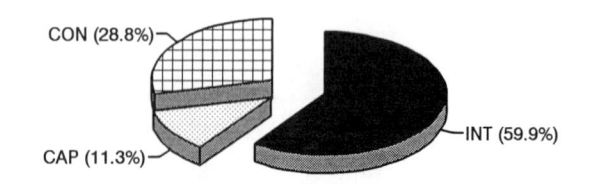

# HONDURAS
## External Trade in US Dollars

### EXPORT STRUCTURE : 1970

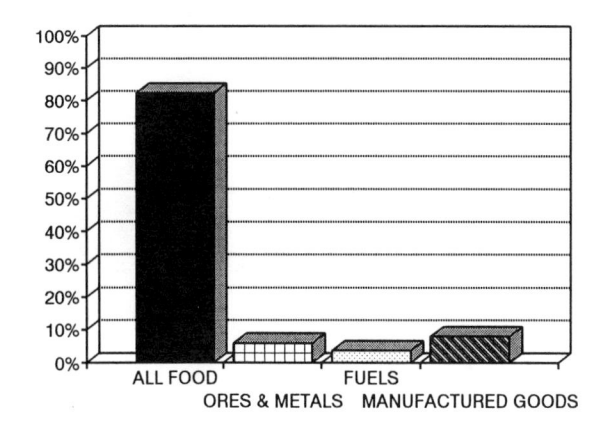

### IMPORT STRUCTURE : 1970

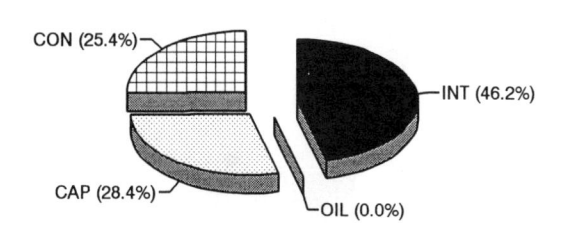

CON (25.4%)
CAP (28.4%)
INT (46.2%)
OIL (0.0%)

### 1980

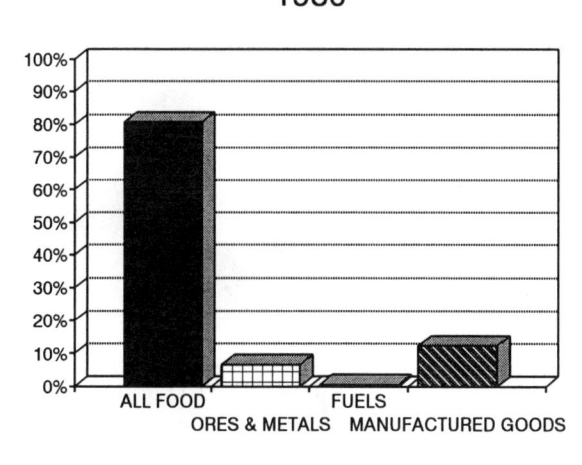

CON (23.3%)
CAP (30.6%)
INT (46.1%)
OIL (0.0%)

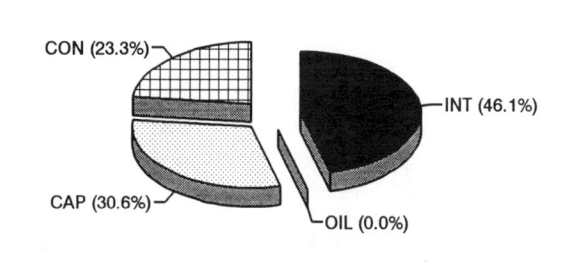

### 1990

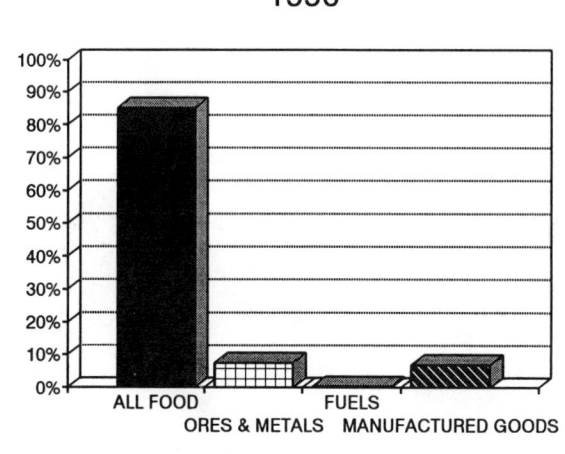

CON (26.6%)
CAP (25.3%)
INT (48.0%)
OIL (0.0%)

Statistics and Quantitative Analysis/IDB

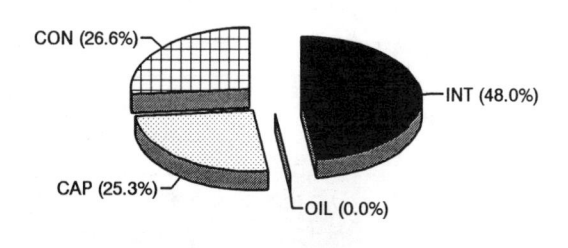

# JAMAICA
## External Trade in US Dollars

### EXPORT STRUCTURE : 1970

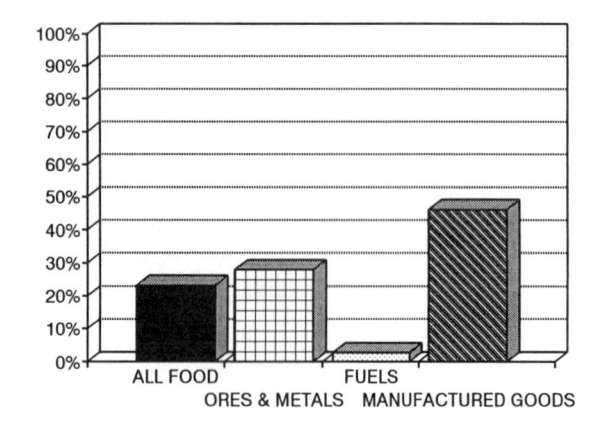

### IMPORT STRUCTURE : 1970

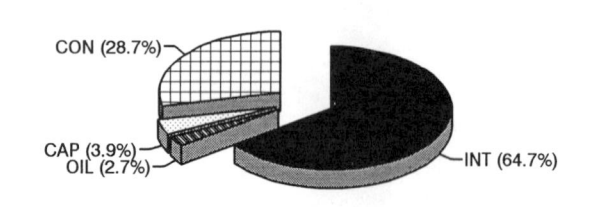

CON (28.7%)
CAP (3.9%)
OIL (2.7%)
INT (64.7%)

### 1980

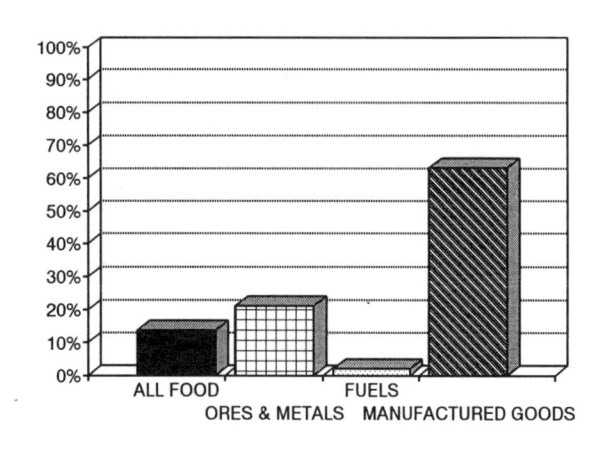

### 1980

CON (11.2%)
CAP (16.8%)
OIL (38.5%)
INT (33.5%)

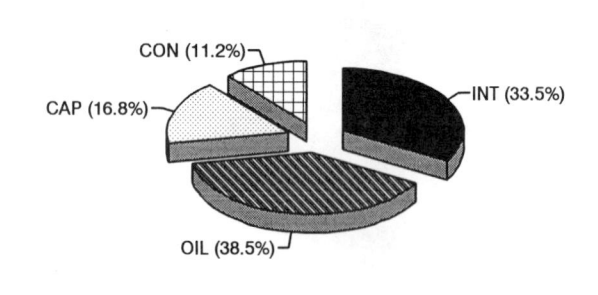

### 1990

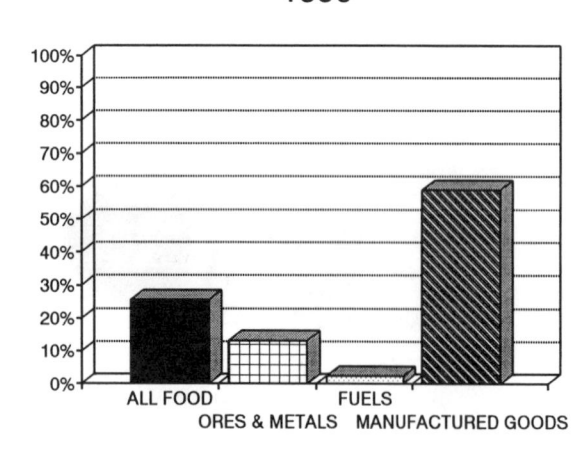

### 1990

CON (17.9%)
CAP (31.0%)
OIL (21.7%)
INT (29.4%)

Statistics and Quantitative Analysis/IDB

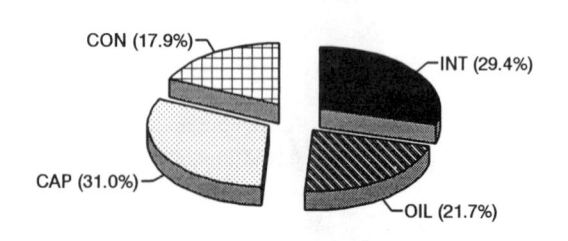

# MEXICO
## External Trade in US Dollars

### EXPORT STRUCTURE: 1970

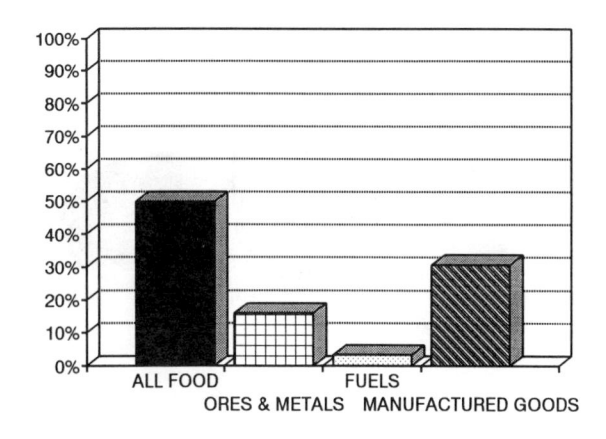

### IMPORT STRUCTURE: 1970

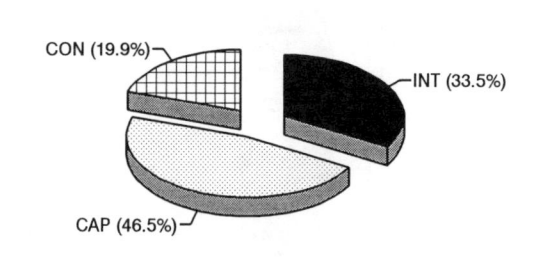

- CON (19.9%)
- INT (33.5%)
- CAP (46.5%)

### 1980

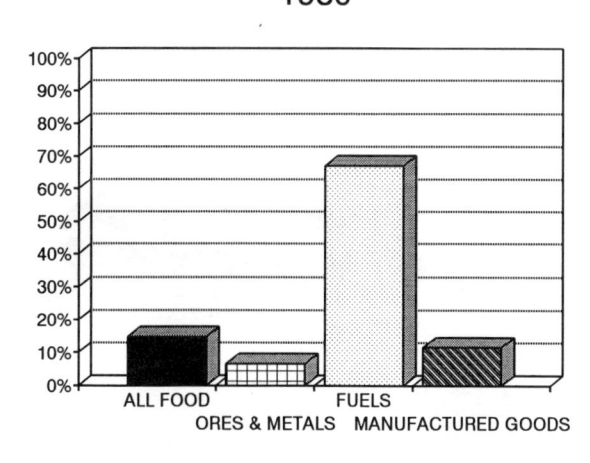

### 1980

- CON (13.1%)
- CAP (27.6%)
- INT (59.4%)

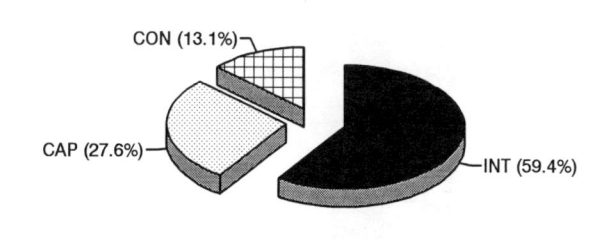

### 1990

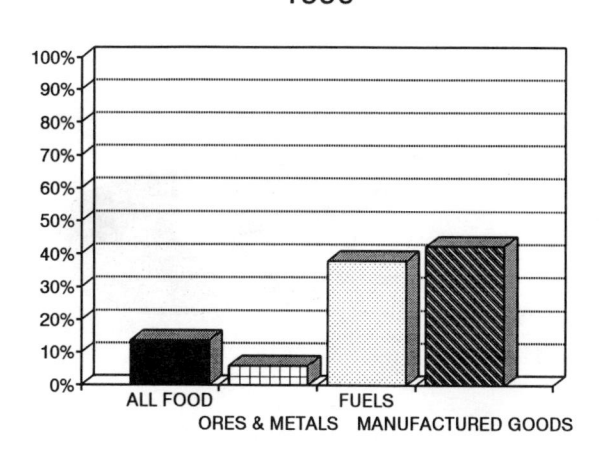

### 1990

- CON (16.3%)
- CAP (21.9%)
- INT (61.8%)

Statistics and Quantitative Analysis/IDB

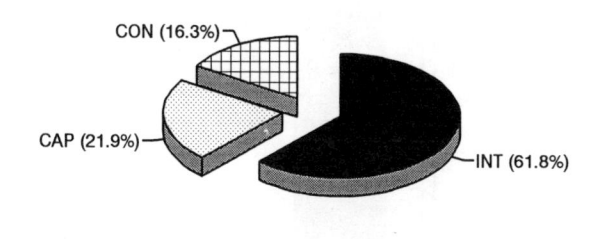

# NICARAGUA
## External Trade in US Dollars

### EXPORT STRUCTURE : 1970

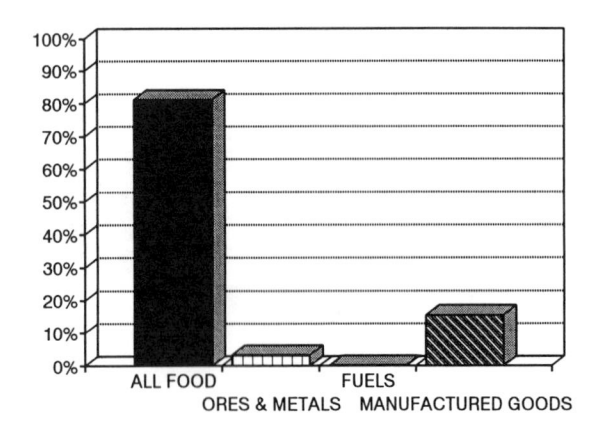

### IMPORT STRUCTURE : 1970

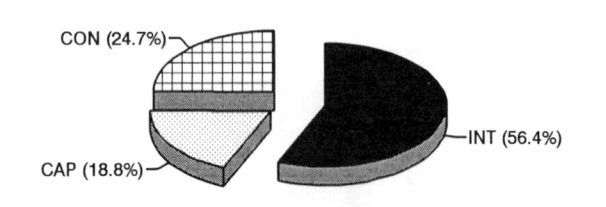

CON (24.7%)
CAP (18.8%)
INT (56.4%)

### 1980

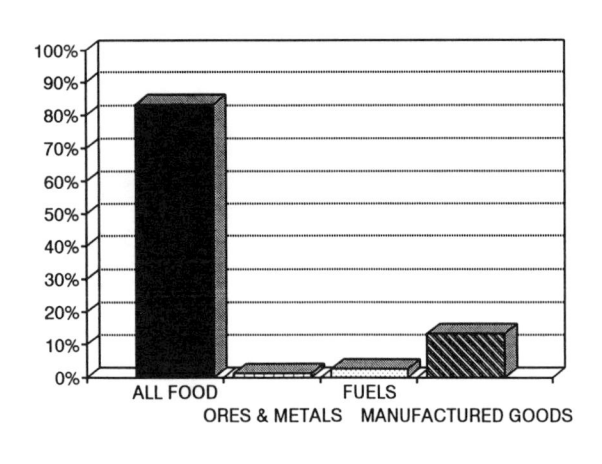

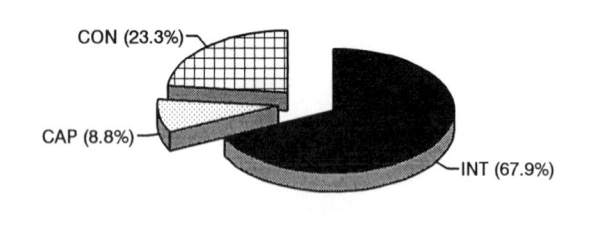

CON (23.3%)
CAP (8.8%)
INT (67.9%)

### 1990

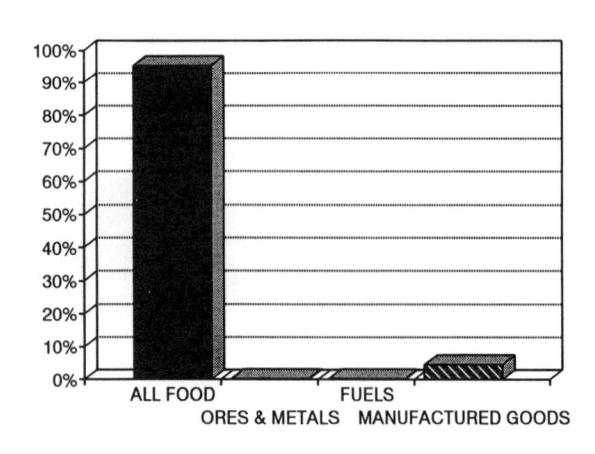

CON (19.0%)
CAP (23.7%)
INT (57.4%)

Statistics and Quantitative Analysis/IDB

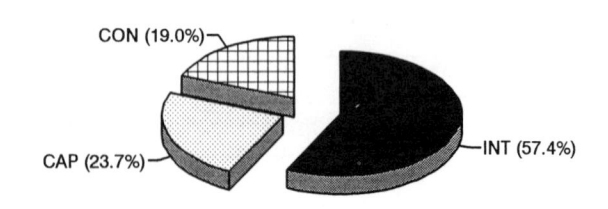

# PANAMA
## External Trade in US Dollars

### EXPORT STRUCTURE : 1970

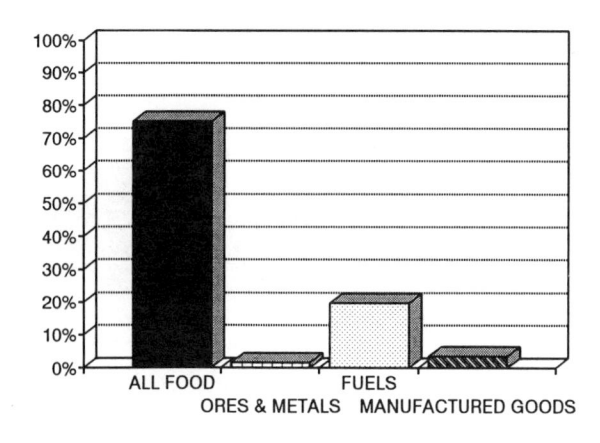

### IMPORT STRUCTURE : 1970

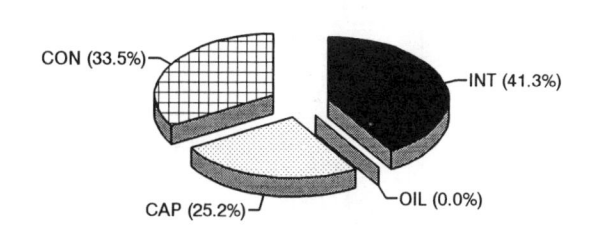

CON (33.5%)
INT (41.3%)
CAP (25.2%)
OIL (0.0%)

### 1980

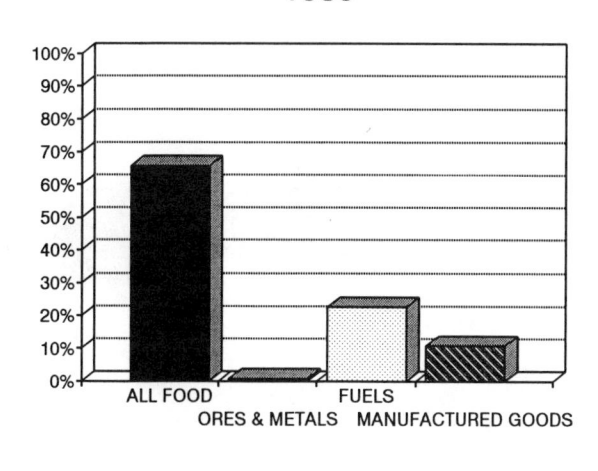

### 1980

CON (34.1%)
INT (42.8%)
CAP (22.0%)
OIL (1.1%)

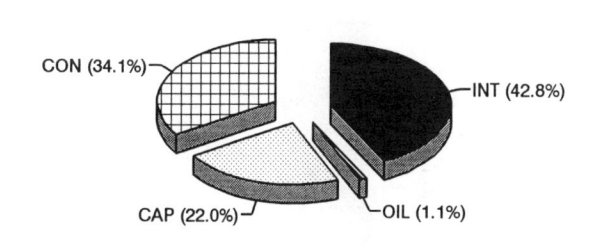

### 1990

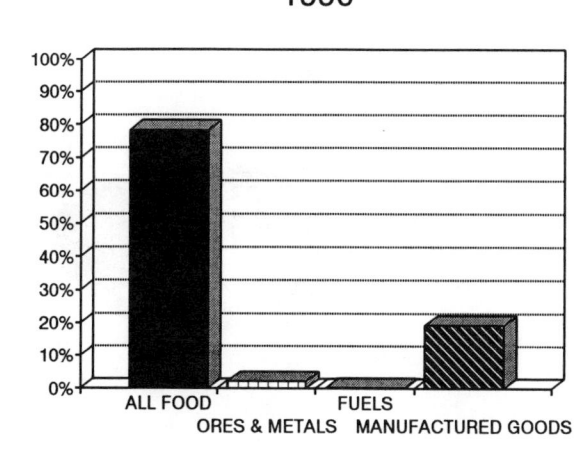

### 1990

CON (28.2%)
CAP (9.2%)
OIL (0.0%)
INT (62.5%)

Statistics and Quantitative Analysis/IDB

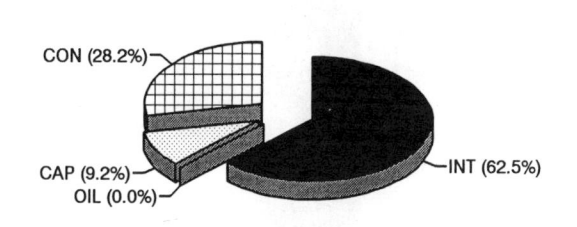

# PARAGUAY
## External Trade in US Dollars

### EXPORT STRUCTURE : 1970

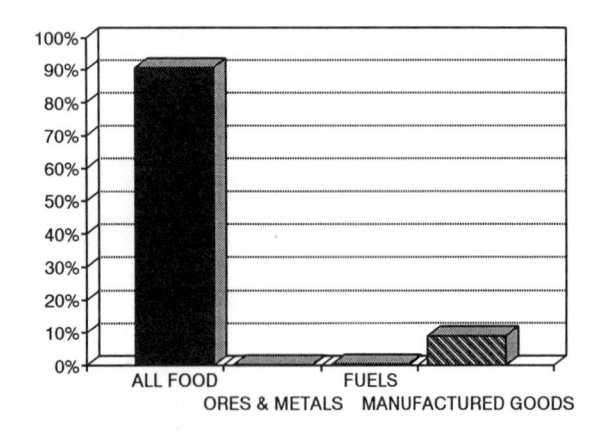

### IMPORT STRUCTURE : 1970

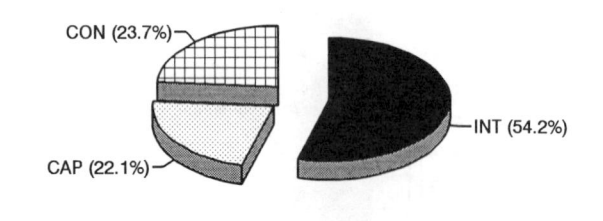

CON (23.7%)
CAP (22.1%)
INT (54.2%)

### 1980

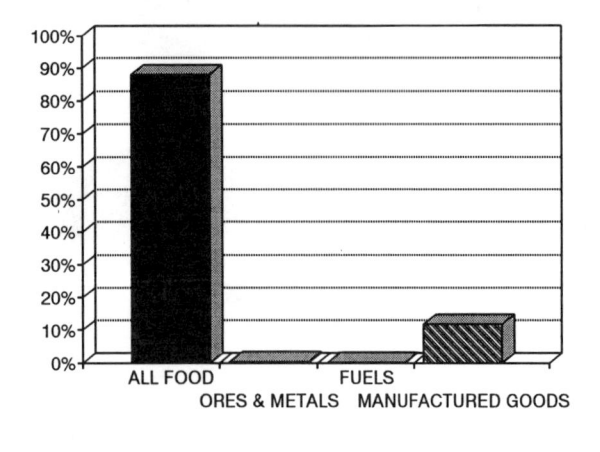

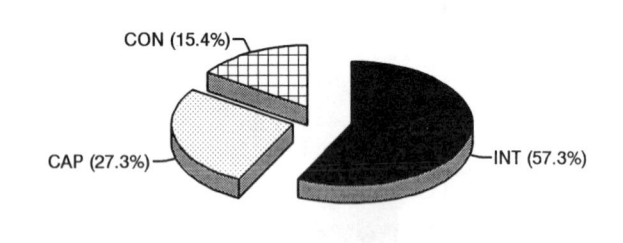

CON (15.4%)
CAP (27.3%)
INT (57.3%)

### 1990

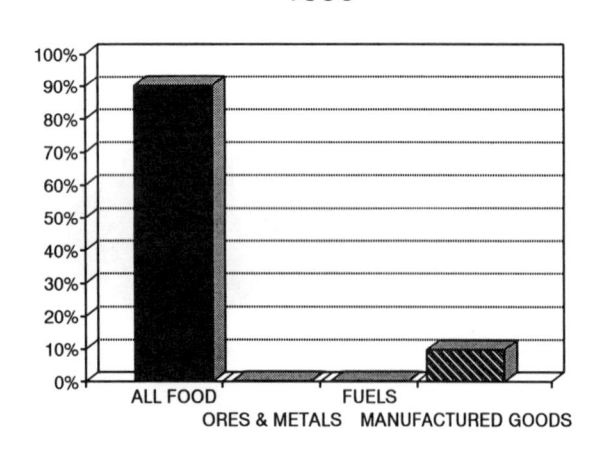

CON (43.8%)
CAP (19.6%)
INT (36.7%)

Statistics and Quantitative Analysis/IDB

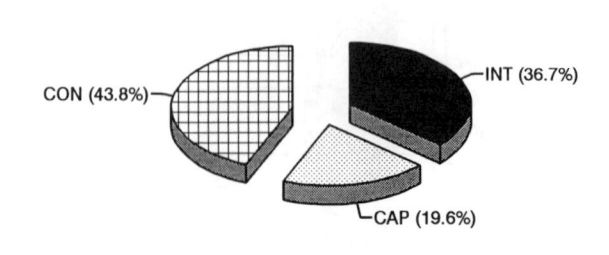

# PERU
## External Trade in US Dollars

### EXPORT STRUCTURE : 1970

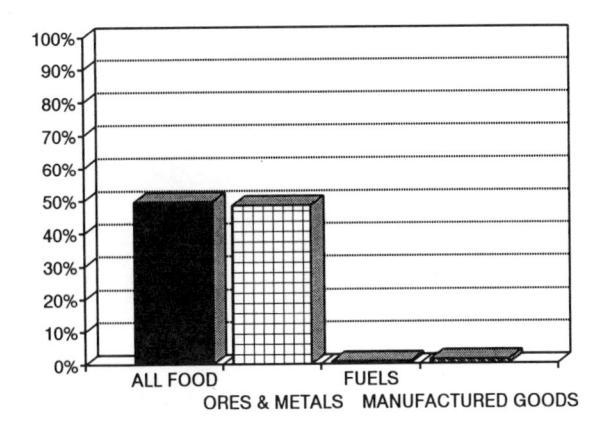

### IMPORT STRUCTURE : 1970
- CON (14.2%)
- CAP (36.1%)
- INT (49.7%)

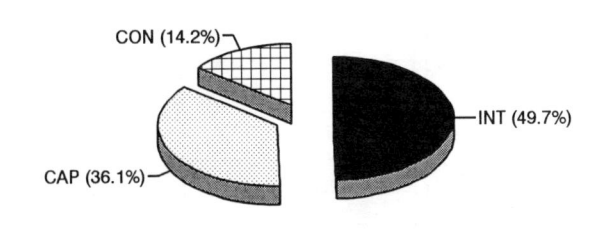

### 1980

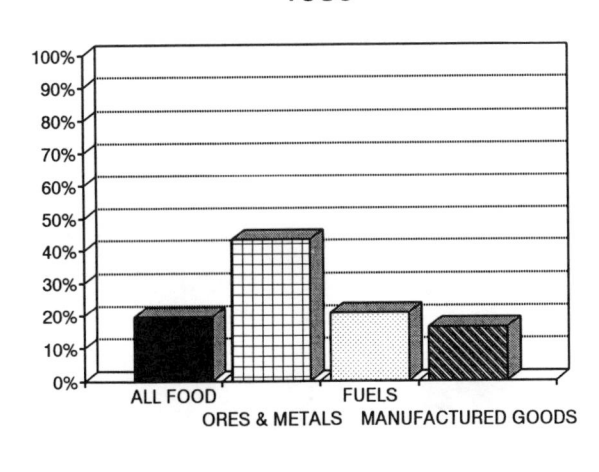

### 1980
- CON (14.4%)
- CAP (42.1%)
- INT (43.5%)

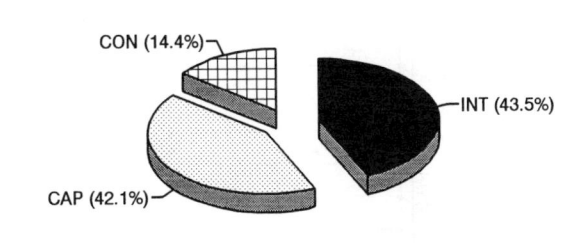

### 1990

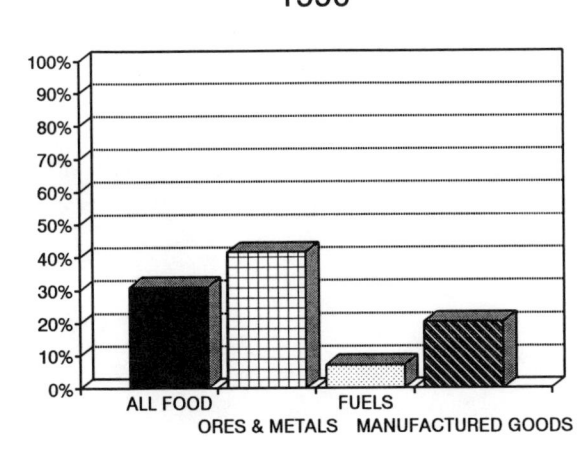

### 1990
- CON (12.9%)
- CAP (36.0%)
- INT (51.1%)

Statistics and Quantitative Analysis/IDB

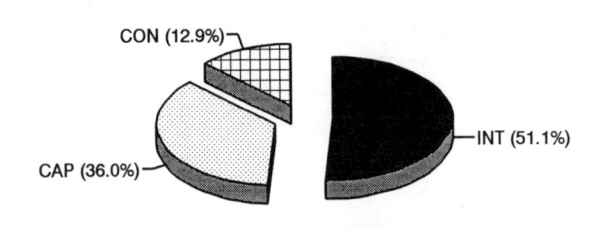

# SURINAME
**External Trade in US Dollars**

### EXPORT STRUCTURE : 1970

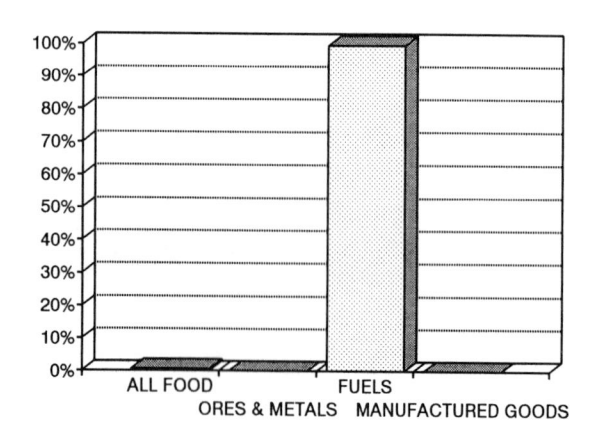

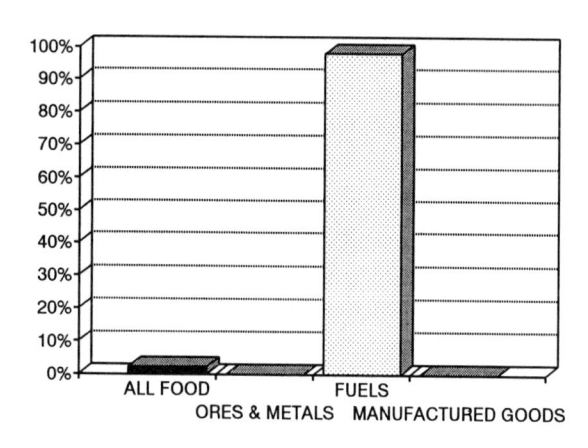

### 1980

### 1990

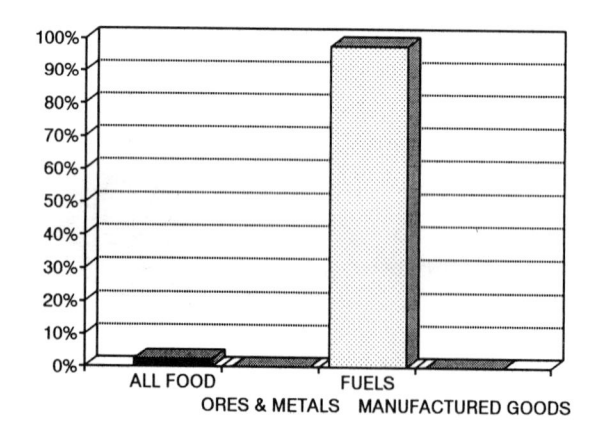

Statistics and Quantitative Analysis/IDB

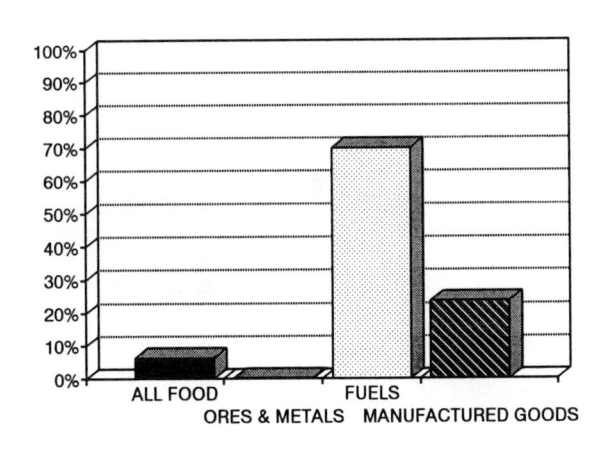

# TRINIDAD AND TOBAGO

**External Trade in US Dollars**

## EXPORT STRUCTURE : 1970

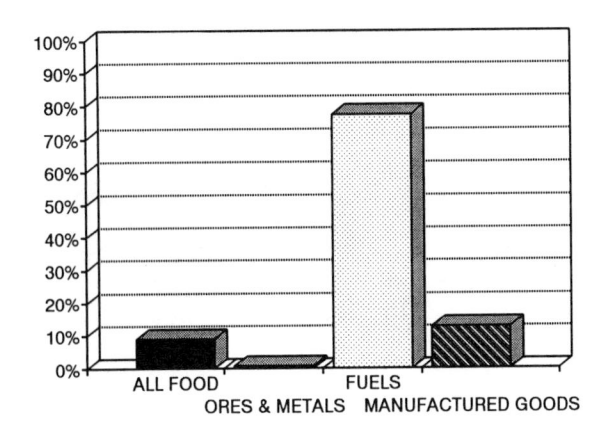

- ALL FOOD: ~7%
- ORES & METALS: ~2%
- FUELS: ~78%
- MANUFACTURED GOODS: ~15%

## IMPORT STRUCTURE : 1970

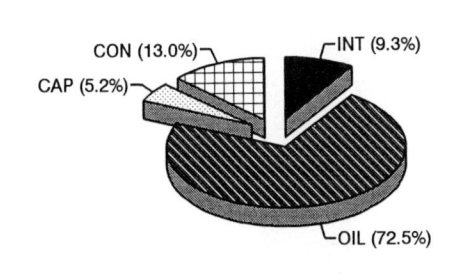

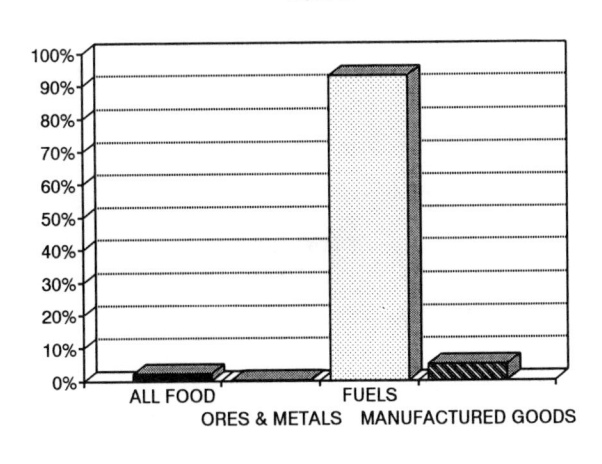

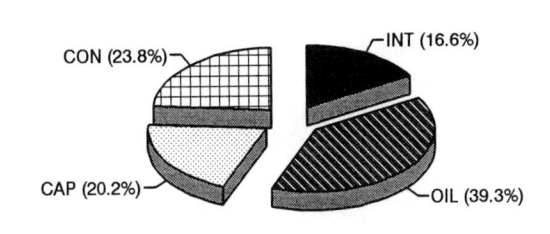

- CON (13.0%)
- CAP (5.2%)
- INT (9.3%)
- OIL (72.5%)

## 1980 (Export)

- ALL FOOD: ~2%
- ORES & METALS: ~1%
- FUELS: ~94%
- MANUFACTURED GOODS: ~6%

## 1980 (Import)

- CON (23.8%)
- CAP (20.2%)
- INT (16.6%)
- OIL (39.3%)

## 1990 (Export)

- ALL FOOD: ~5%
- ORES & METALS: ~1%
- FUELS: ~70%
- MANUFACTURED GOODS: ~24%

## 1990 (Import)

- CON (32.4%)
- CAP (21.7%)
- INT (33.5%)
- OIL (12.3%)

Statistics and Quantitative Analysis/IDB

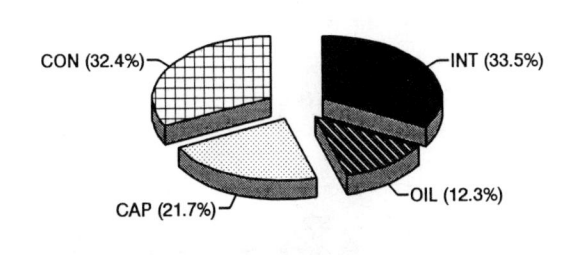

# URUGUAY
## External Trade in US Dollars

### EXPORT STRUCTURE : 1970

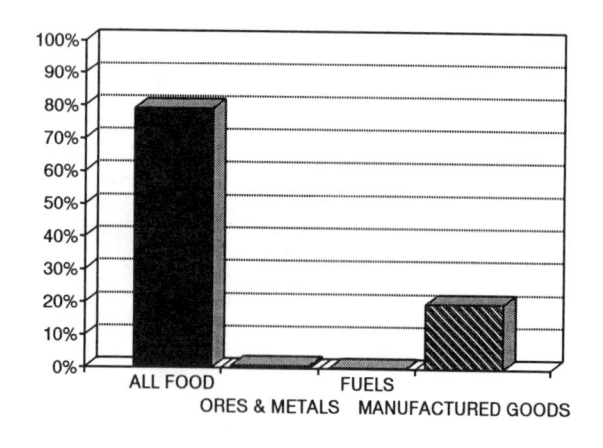

### IMPORT STRUCTURE : 1970
- CON (10.1%)
- CAP (16.6%)
- OIL (14.6%)
- INT (58.8%)

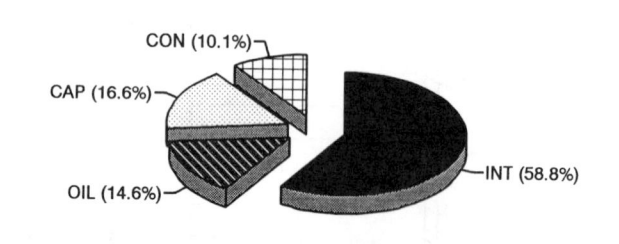

### 1980

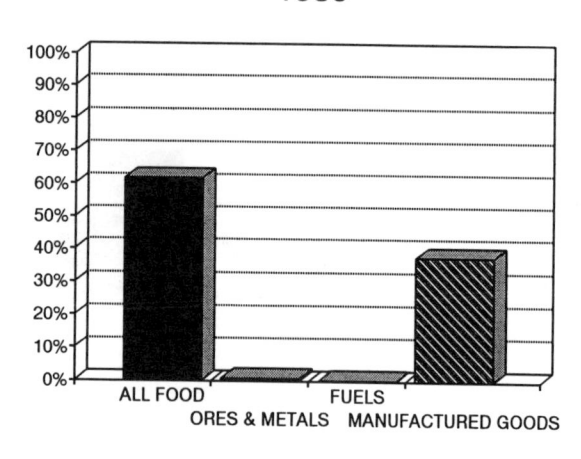

### 1980
- CON (10.6%)
- CAP (16.4%)
- OIL (27.1%)
- INT (45.9%)

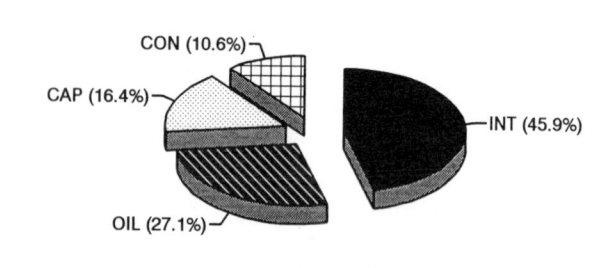

### 1990

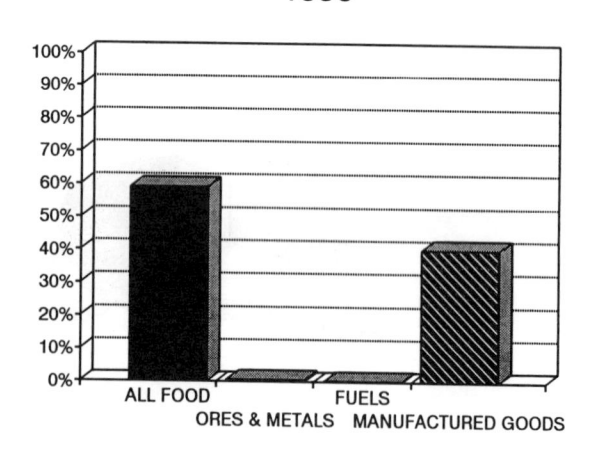

### 1990
- CON (14.9%)
- CAP (13.3%)
- OIL (15.1%)
- INT (56.7%)

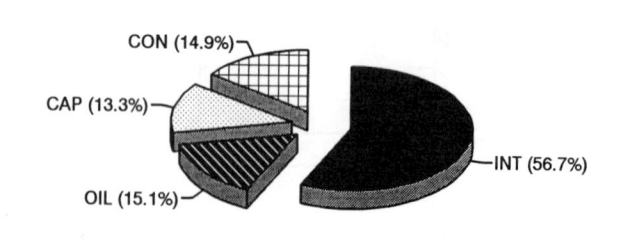

# VENEZUELA
## External Trade in US Dollars

### EXPORT STRUCTURE : 1970

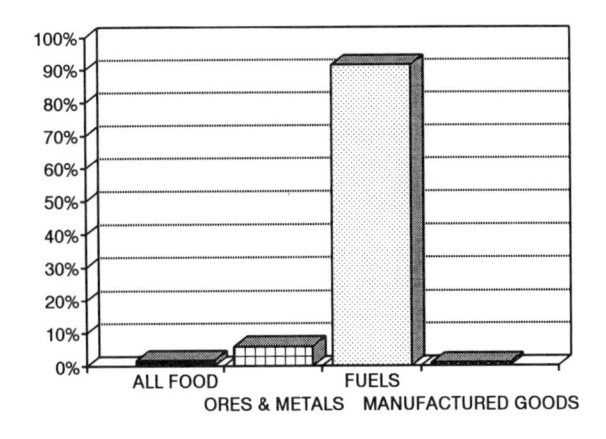

### IMPORT STRUCTURE : 1970

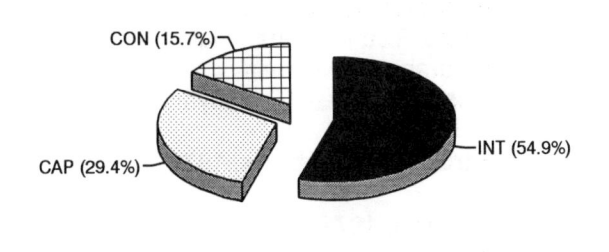

CON (15.7%)
CAP (29.4%)
INT (54.9%)

### 1980

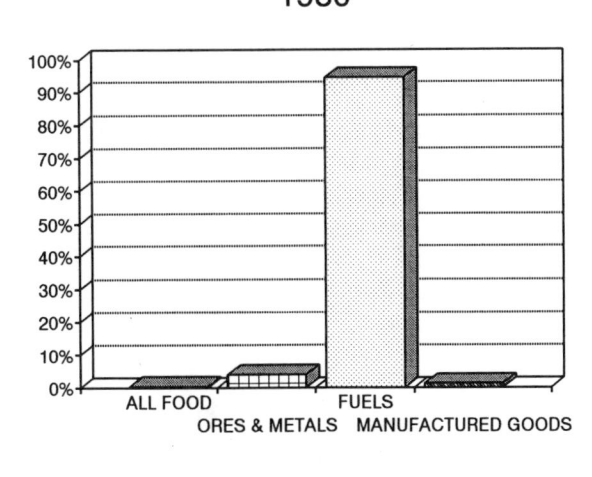

### 1980

CON (18.8%)
CAP (30.5%)
INT (50.8%)

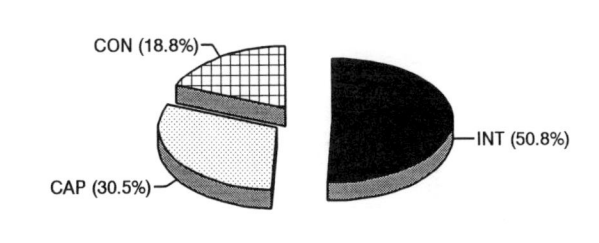

### 1990

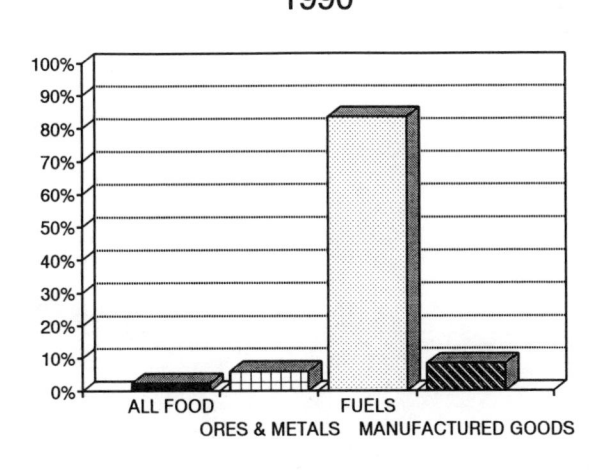

### 1990

CON (11.5%)
CAP (30.5%)
INT (58.0%)

Statistics and Quantitative Analysis/IDB

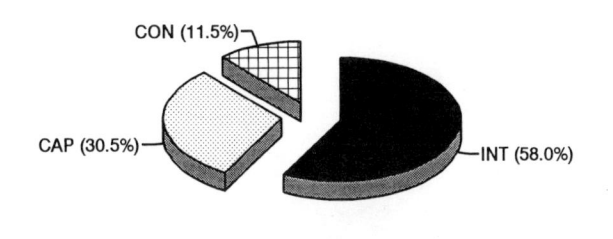

## Country Notes on External Trade

Latin America excludes Bahamas and Suriname. Most proximate years for Export and Import structures are:

### Structure of Exports: 1990

Ecuador, Trinidad and Tobago and Uruguay, 1991;
Honduras, Panama and Peru, 1989;
Bahamas and Jamaica, 1988;
Costa Rica and Guatemala, 1987;
Nicaragua, 1986; and
Dominican Republic, Guyana, Haiti and Suriname, 1983.

### Structure of Imports: 1970

Jamaica, 1971;
Trinidad and Tobago, 1974;
Dominican Republic, 1975; and
Haiti, 1977.

### Structure of Imports: 1990

Brazil, Costa Rica, Guatemala, Mexico, Nicaragua, Panama and Peru, 1989;
Honduras, 1988; and
Dominican Republic, 1985.

# EXTERNAL DEBT

# EXTERNAL DEBT

Data on Debt are from the World Bank, World Debt Tables tapes and World Bank estimates.

## DISBURSED EXTERNAL DEBT AS A PERCENT OF GDP

The graph shows the evolution of the structure of Total External Debt disbursed as a percent of GDP in current Dollars at year's end. *Long-Term Debt* has an original or extended maturity of more than one year. *Short-Term Debt* includes the stock of Interest in Arrears on Long-Term Debt. The *Use of IMF Credit* includes purchases outstanding under the Credit Tranche, Special Facilities (e.g., the Oil Facility), Trust Fund loans and the Structural Adjustment loans.

Data on *Short-Term Debt* are not available prior to 1977 and have been estimated for previous years by applying the ratio of *Short-Term Debt* to *Long-Term Debt* in 1977. In most countries, the close relationship between the evolution of External Debt and the accumulated Current Account deficits, as shown in another of the graphs, is indicative of the rather good quality of these estimates.

## STRUCTURE OF DISBURSED EXTERNAL DEBT, END 1991

The graph gives the structure of Total External Debt disbursed by type of creditor. *Multilateral Debt* includes debt from Multilateral Institutions as well as the *Use of IMF Credit*. *Bilateral Debt* mainly includes loans from Governments and their agencies (including official export credit agencies). *Private Debt* includes *Private Long-Term Non-guaranteed Debt*, *Public Long-Term Guaranteed Debt* other than Bilateral or Multilateral (mainly bonds and exports credits) and *Short-Term Debt*.

## MULTILATERAL AS A PERCENT OF TOTAL DISBURSED DEBT

The graph shows the proportion of IMF Credits and Multilateral credits outstanding (mainly from the IDB and the World Bank) in Total External Debt disbursed.

## INTEREST PAYMENTS AS A PERCENT OF EXPORTS

This ratio shows the proportion of earnings from Exports of Goods and Non-factor Services which has to be dedicated to interest payments which are due (accrued basis) according to balance of payments figures. It also shows the proportion which was actually dedicated to interest payments (actually paid), as recorded by the World Bank.

## IMPLICIT INTEREST RATE VERSUS LIBOR

The implicit interest rate is the ratio of accrued interest payments over the debt stock at mid-year (estimated as the simple average of start and end year stocks of Total External Debt disbursed). The three-month LIBOR (London InterBank Offered Rate) is taken as an indicator of the international level of interest rates. The gap depends on the respective proportion of concessional and variable interest rates in Total External Debt disbursed, as well as the average grace period and the spread.

## DEBT AND ACCUMULATED CURRENT DEFICITS

*External Debt* is Total External Debt disbursed. *Net External Debt* is the External Debt net of the Stock of Reserves of the country (excluding gold). *Net External Debt* should theoretically match the accumulation of Current Account deficits in the case where net external borrowing is the unique source of financing.

The *Accumulated deficits* series are the simple accumulation of Current Account deficits added to the initial debt at the end of 1971. Nevertheless, these series generally do not match the *Net External Debt* series because of the impact of inflation and exchange rate movements, because a part of the deficit may be financed by Net Direct and Portfolio Investment, and because of Errors and Omissions. However, for some countries, the main explanation for the increasing gap in the early 1980's is capital flight. In more recent years, the gap has been reduced by debt reduction, return of capital and Net Direct and Portfolio Investment.

# LATIN AMERICA
## External Debt in US Dollars

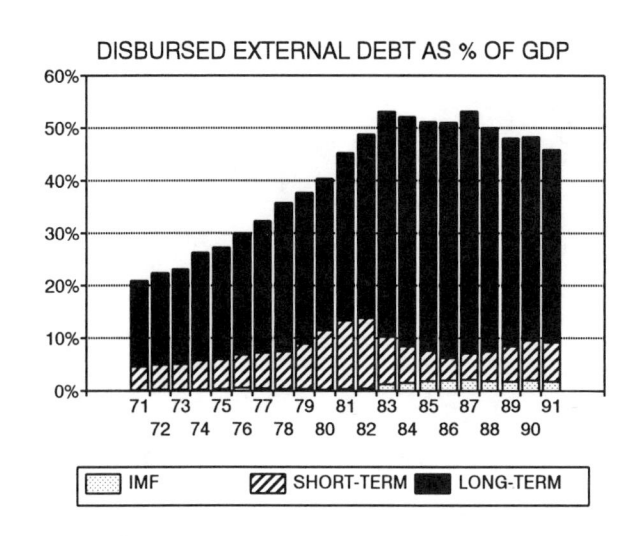

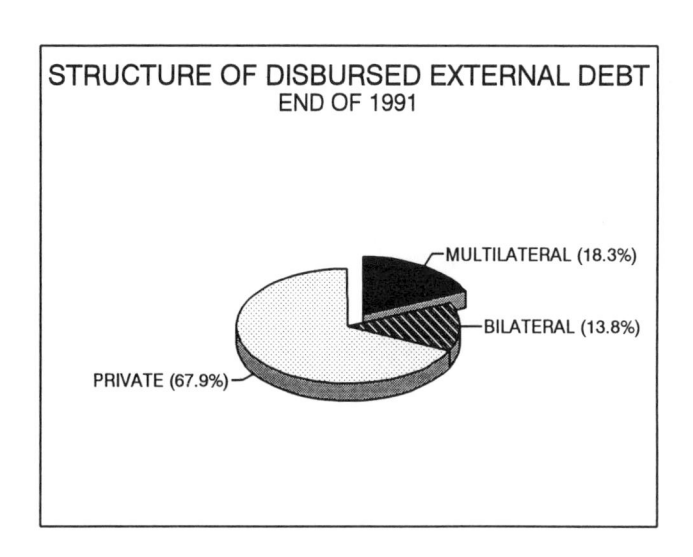

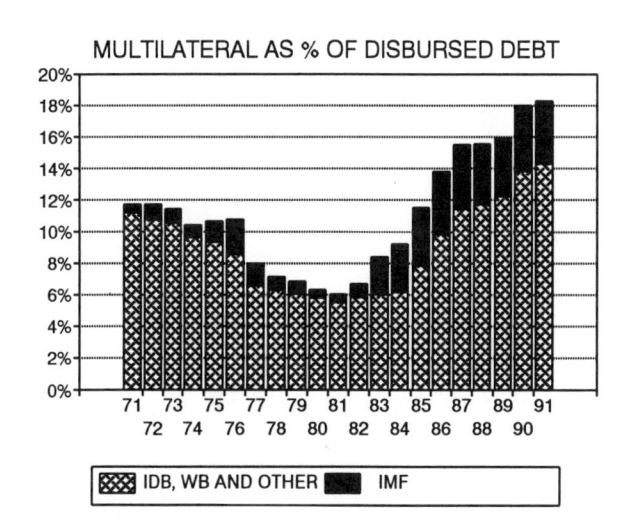

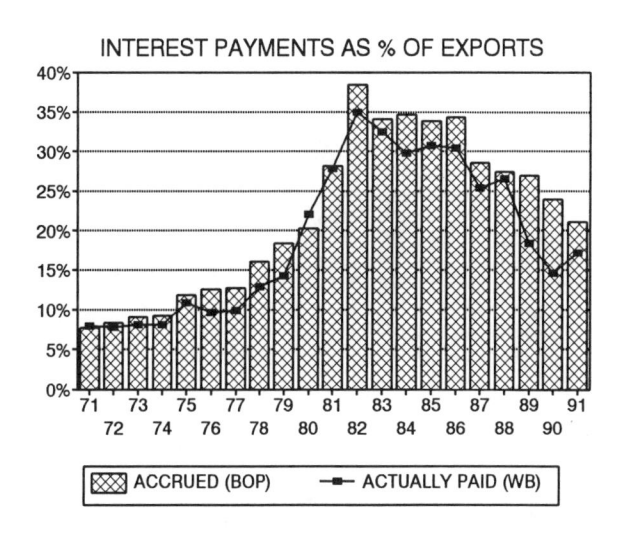

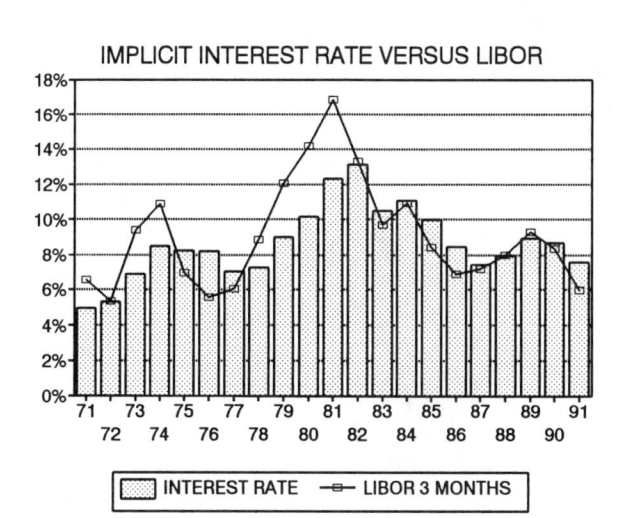

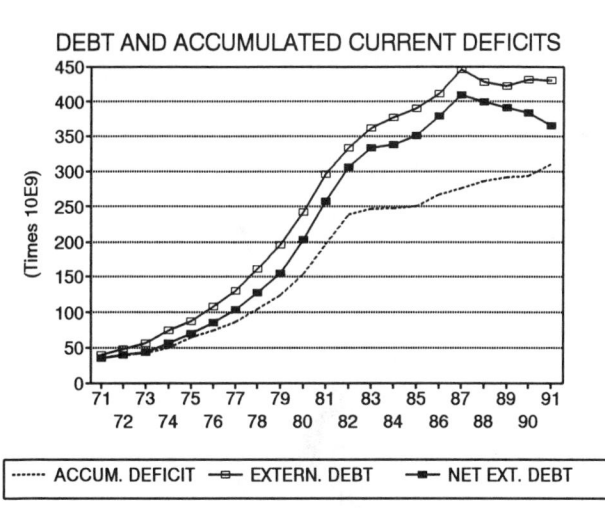

### DISBURSED EXTERNAL DEBT AS % OF GDP
IMF / SHORT-TERM / LONG-TERM

### STRUCTURE OF DISBURSED EXTERNAL DEBT
END OF 1991

- MULTILATERAL (18.3%)
- BILATERAL (13.8%)
- PRIVATE (67.9%)

### MULTILATERAL AS % OF DISBURSED DEBT
IDB, WB AND OTHER / IMF

### INTEREST PAYMENTS AS % OF EXPORTS
ACCRUED (BOP) / ACTUALLY PAID (WB)

### IMPLICIT INTEREST RATE VERSUS LIBOR
INTEREST RATE / LIBOR 3 MONTHS

### DEBT AND ACCUMULATED CURRENT DEFICITS
(Times 10E9)
ACCUM. DEFICIT / EXTERN. DEBT / NET EXT. DEBT

Statistics and Quantitative Analysis/IDB

# ARGENTINA
## External Debt in US Dollars

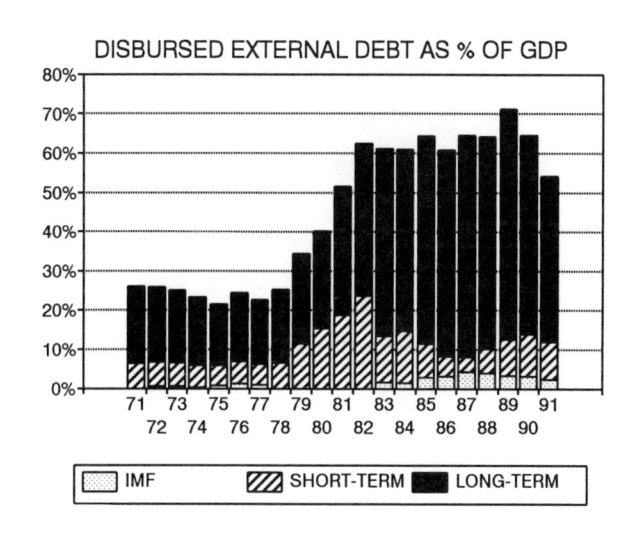

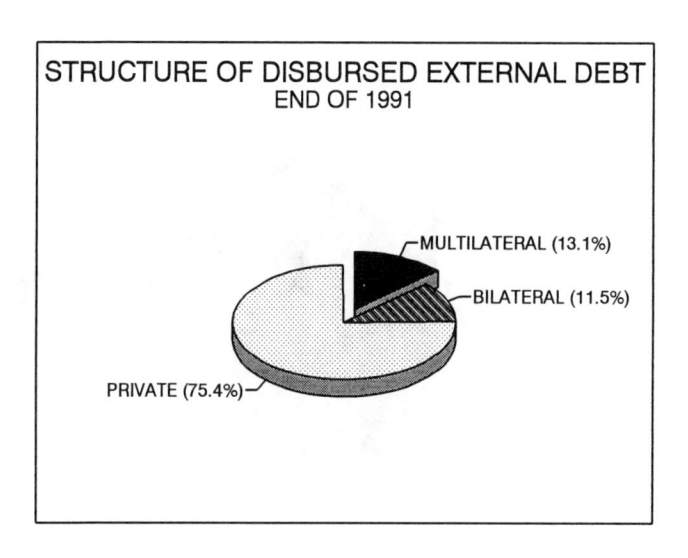

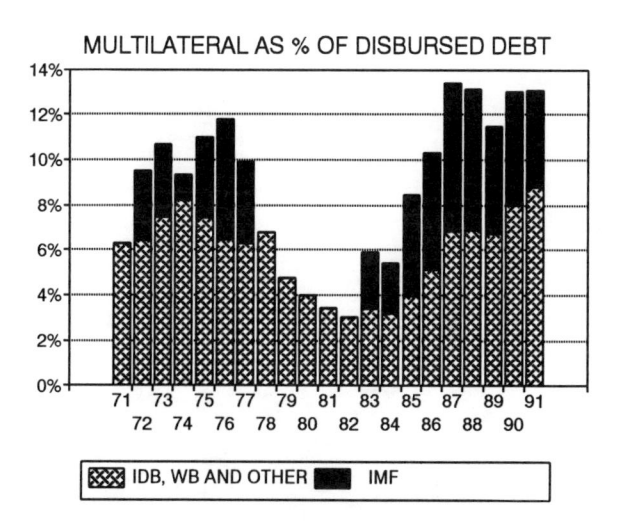

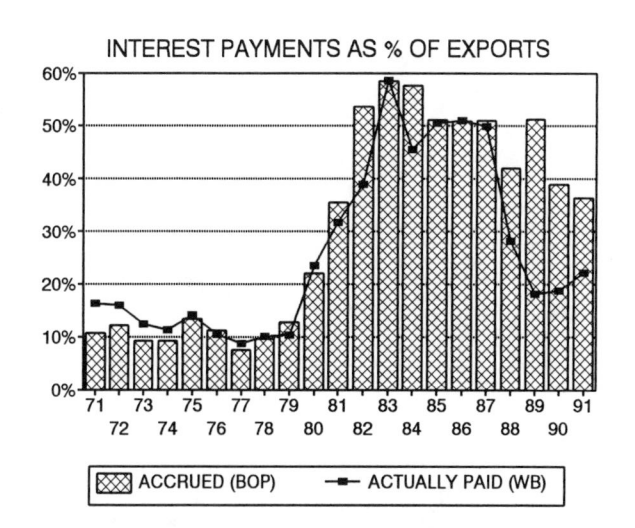

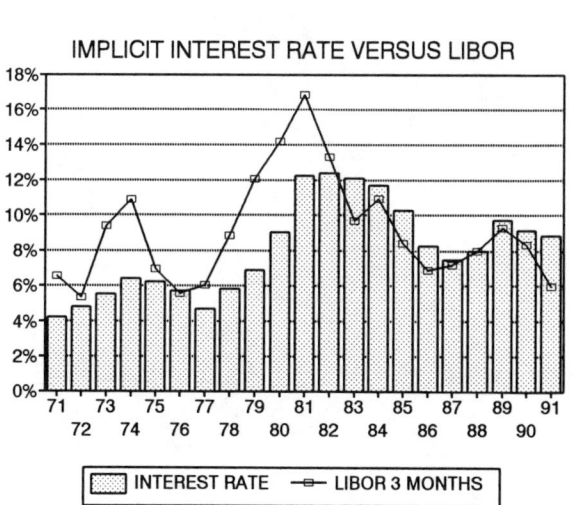

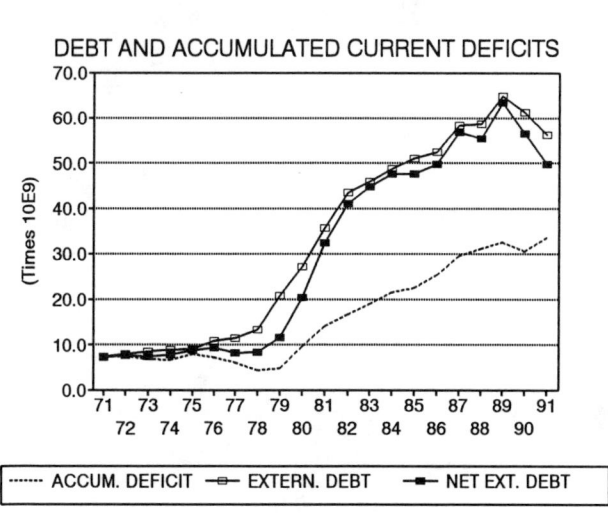

Statistics and Quantitative Analysis/IDB

103

# BAHAMAS
## External Debt in US Dollars

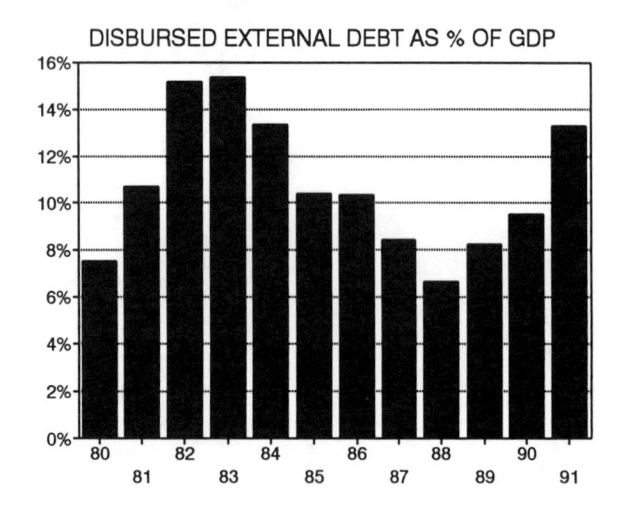

DISBURSED EXTERNAL DEBT AS % OF GDP

# BARBADOS
## External Debt in US Dollars

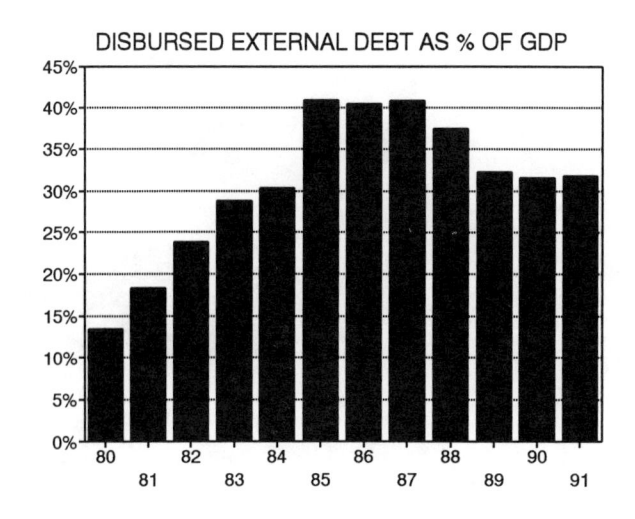

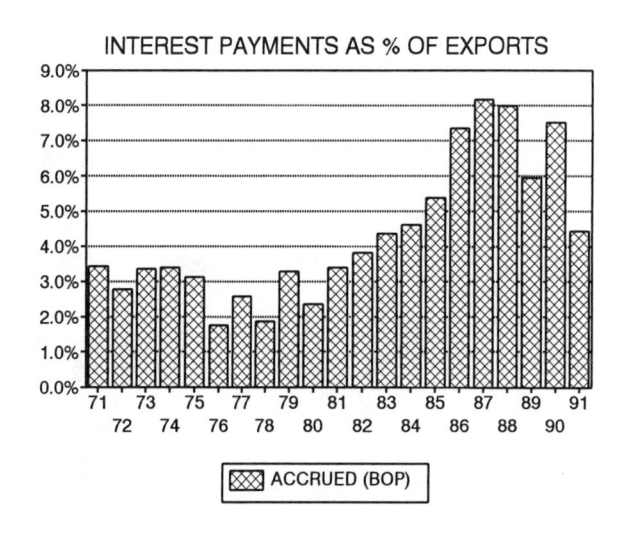

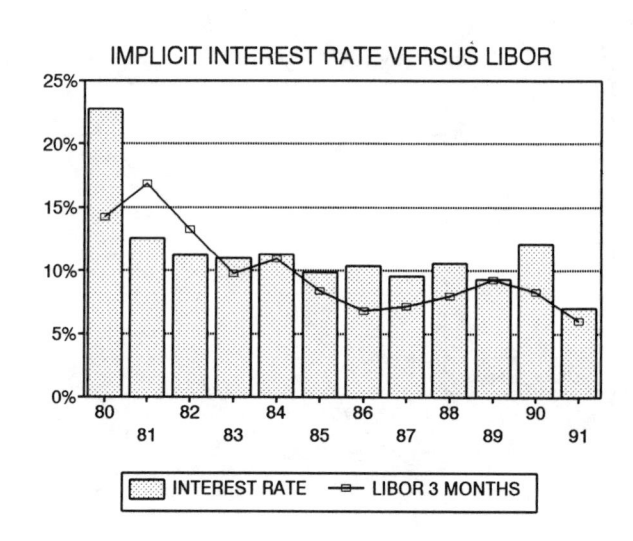

Statistics and Quantitative Analysis/IDB

105

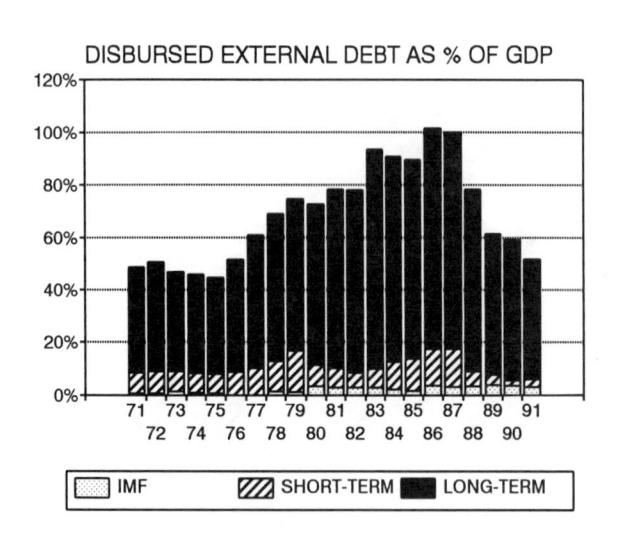

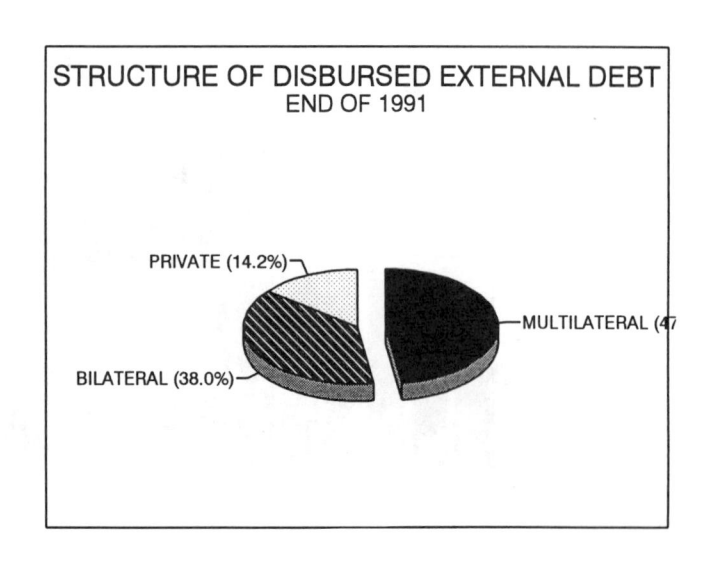

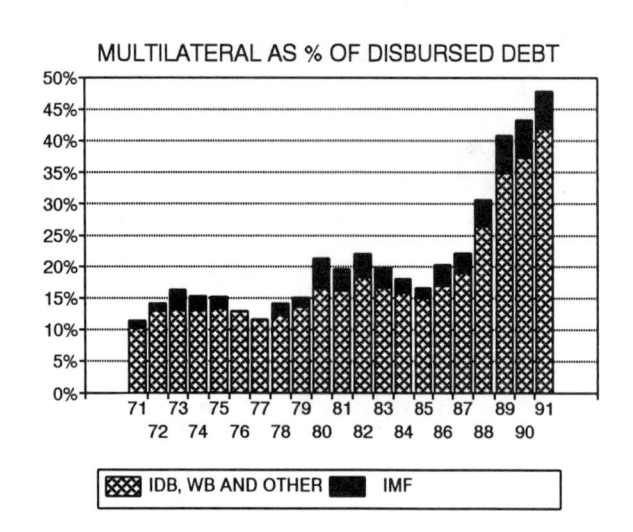

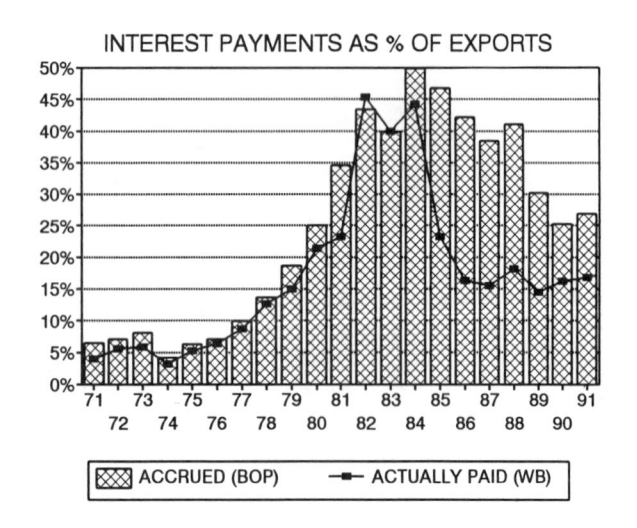

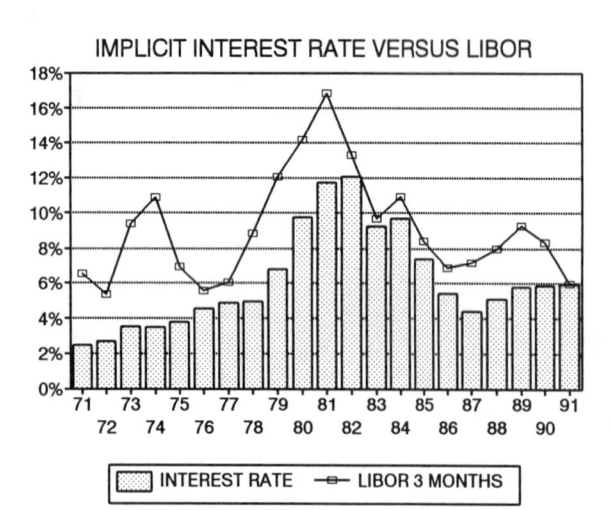

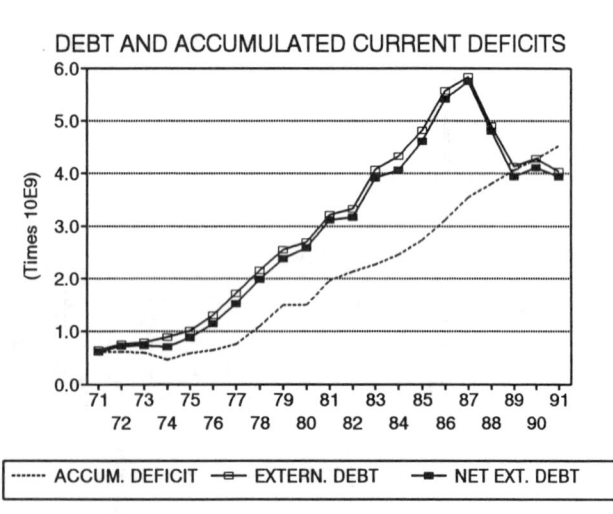

# BOLIVIA
## External Debt in US Dollars

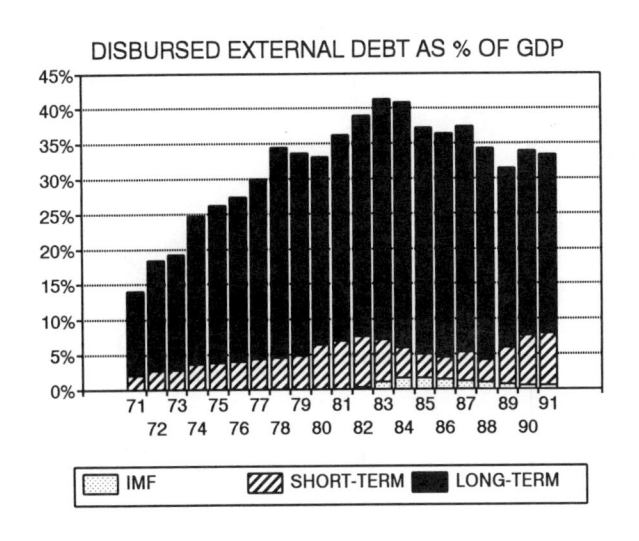

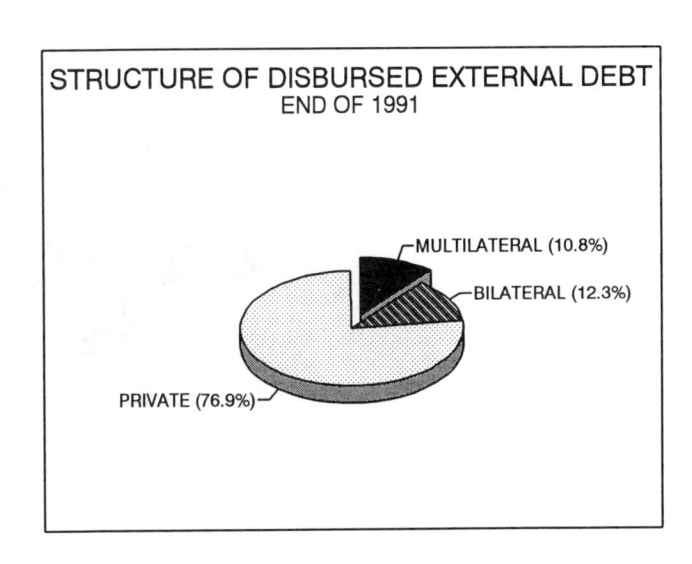

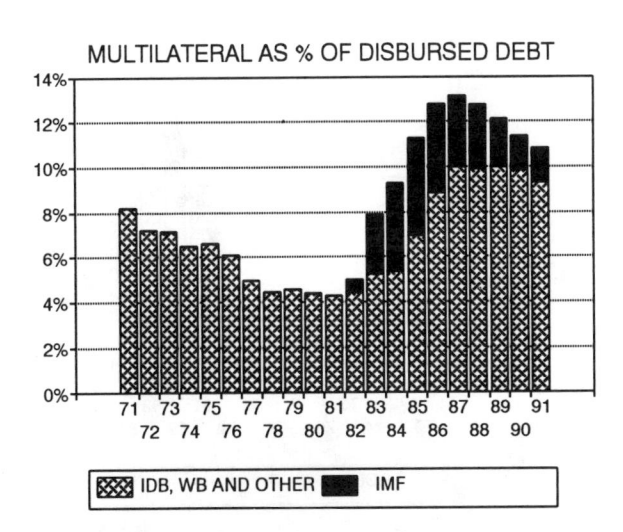

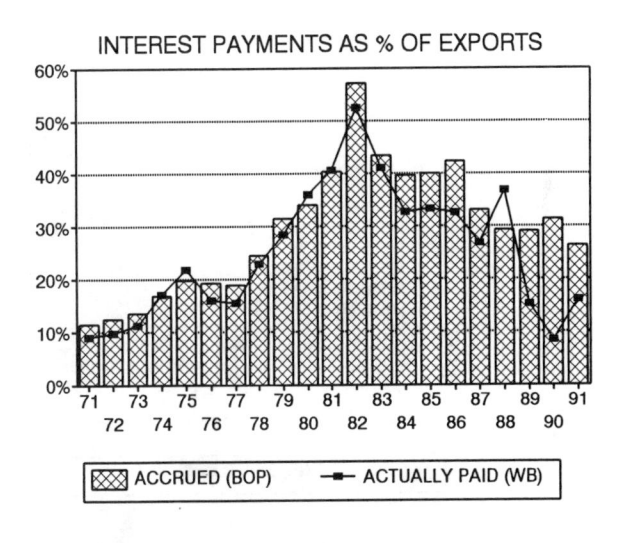

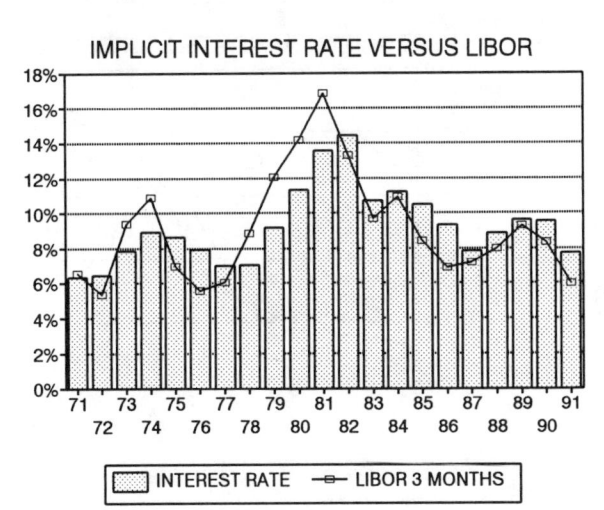

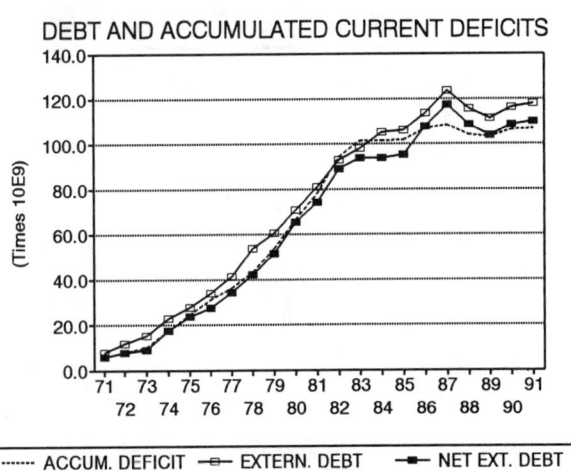

# BRAZIL
## External Debt in US Dollars

Statistics and Quantitative Analysis/IDB

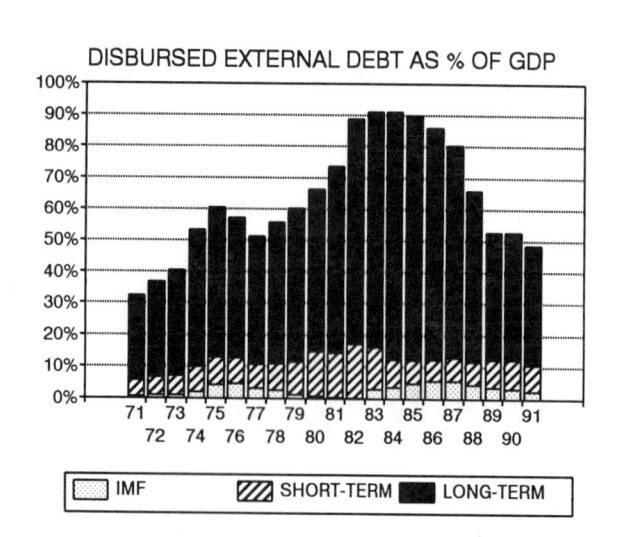

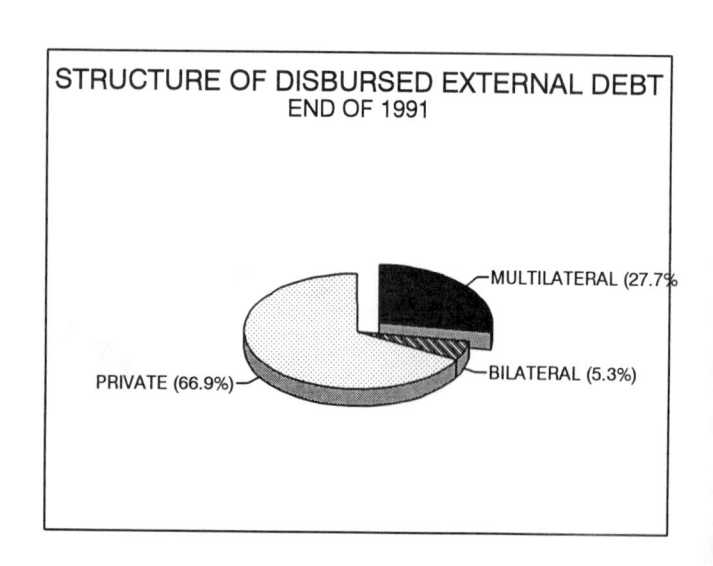

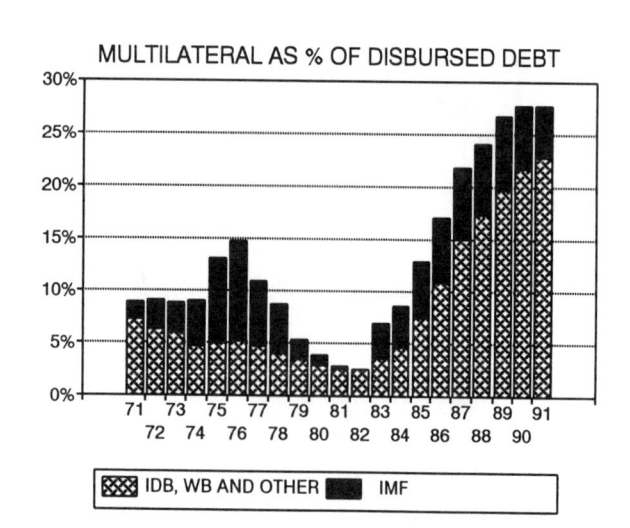

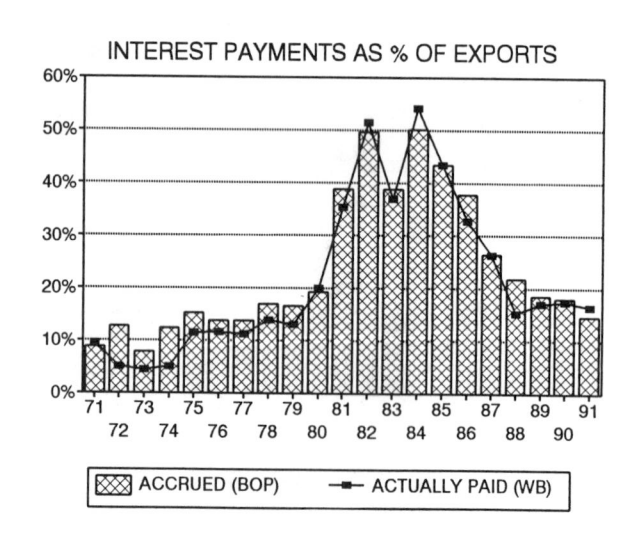

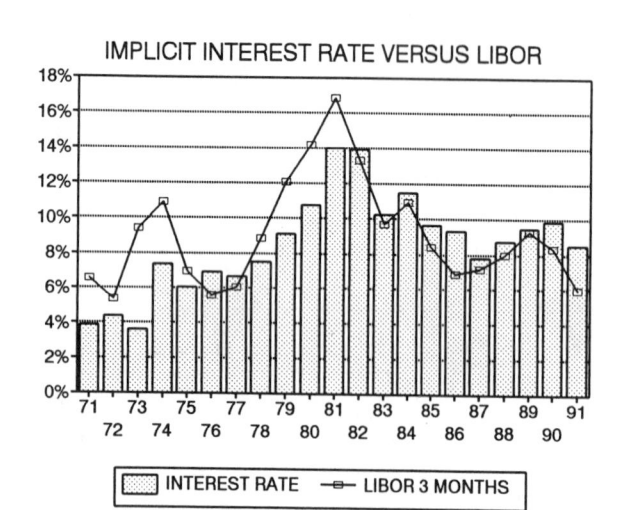

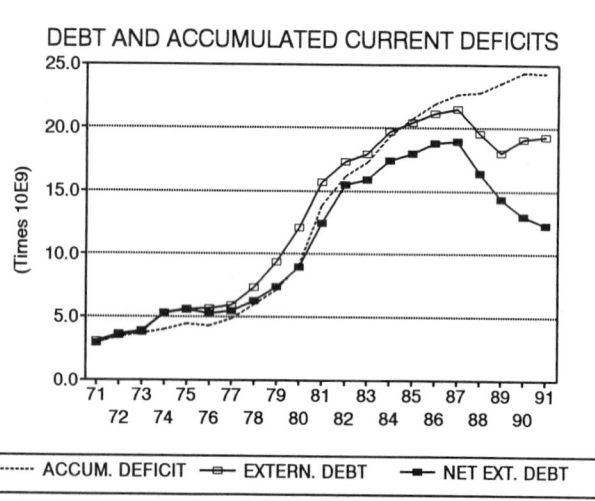

# CHILE
## External Debt in US Dollars

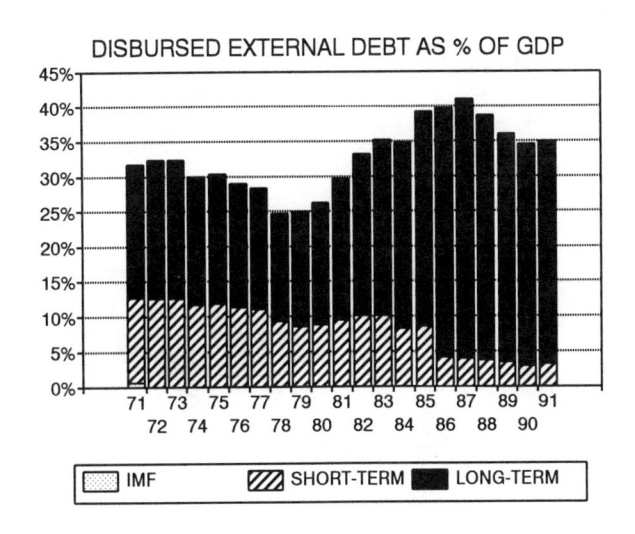

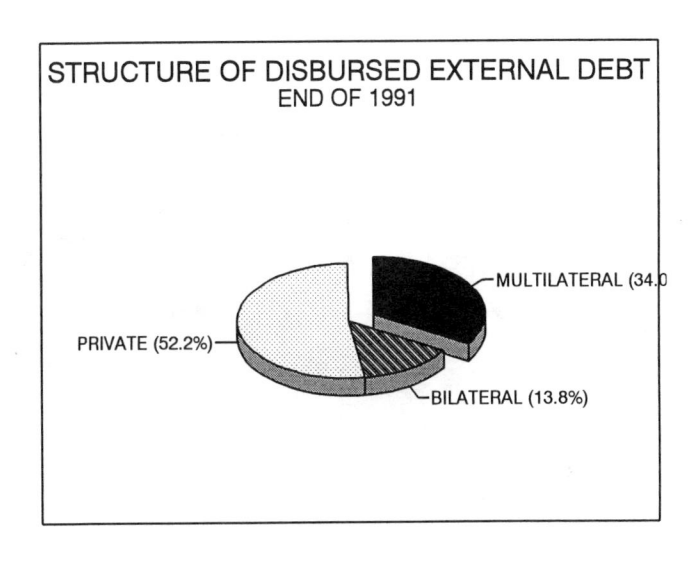

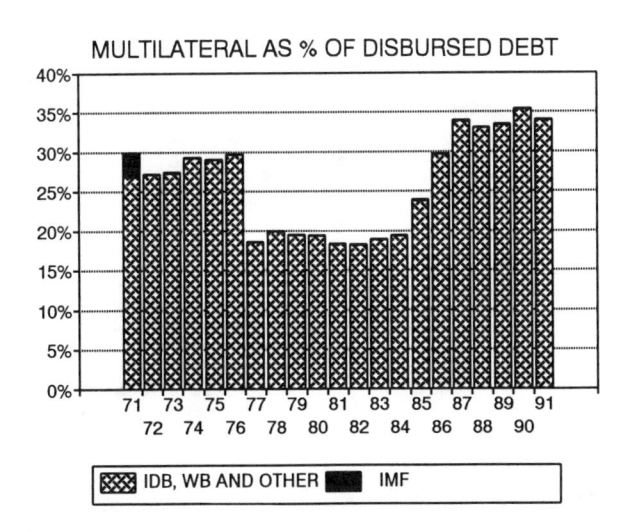

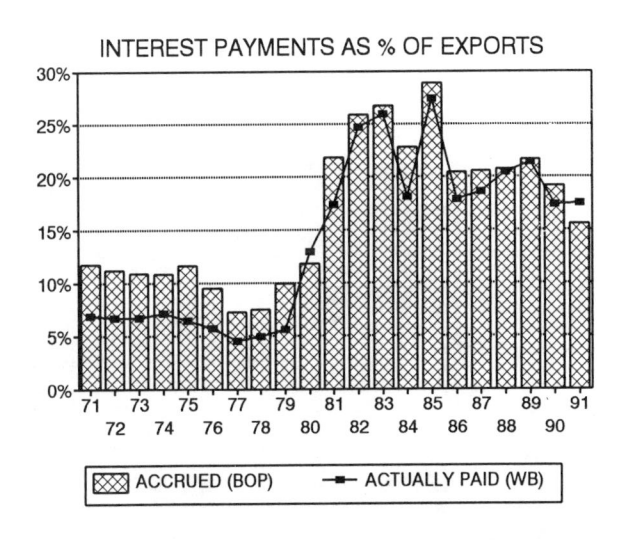

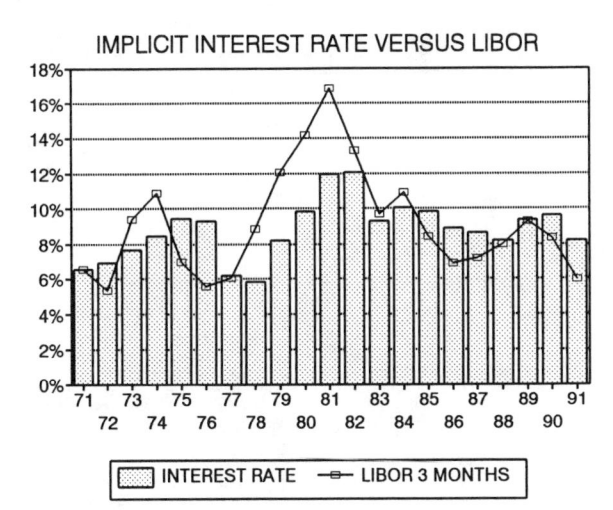

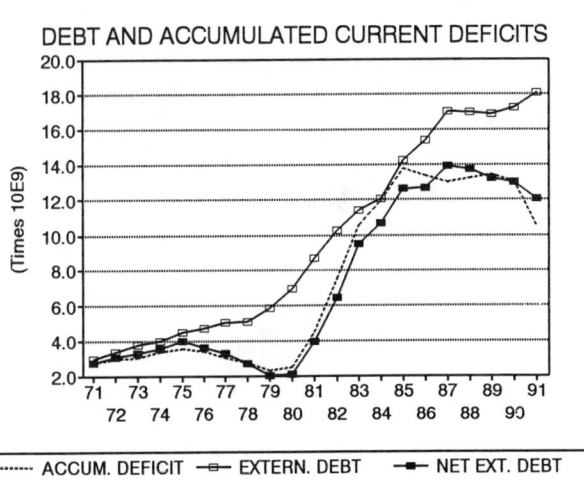

# COLOMBIA
## External Debt in US Dollars

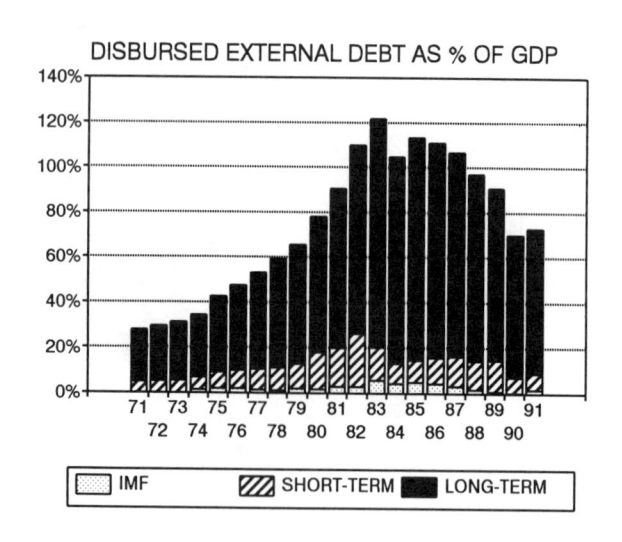

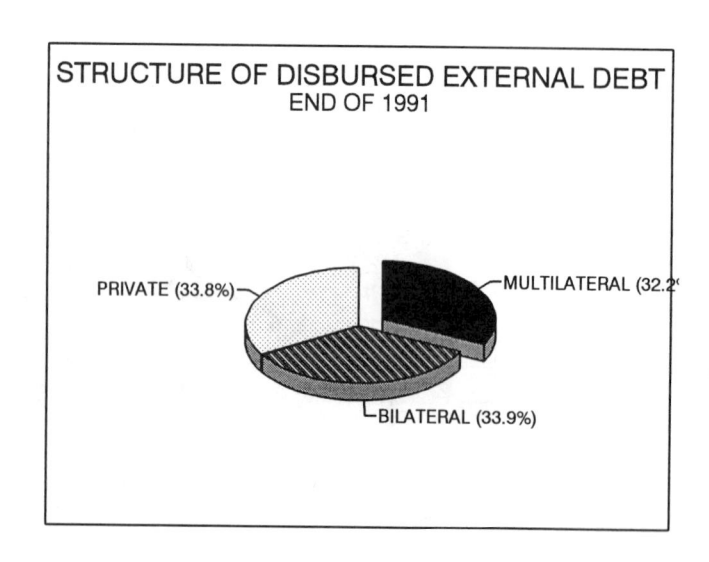

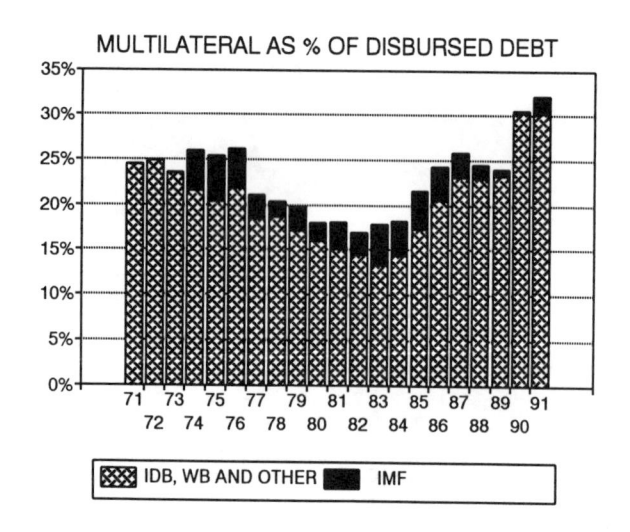

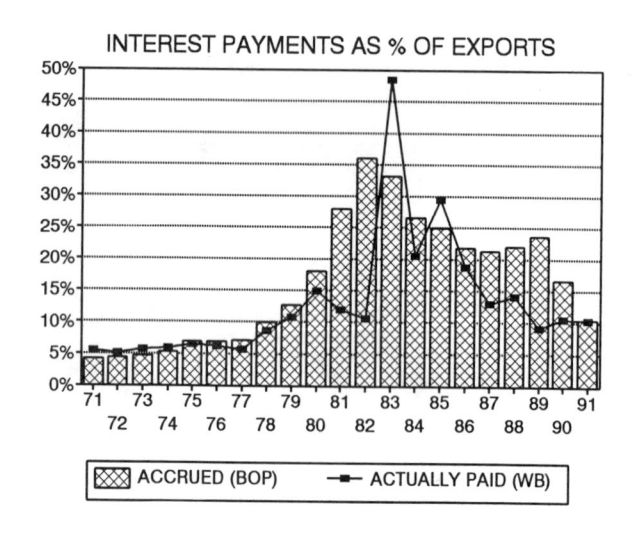

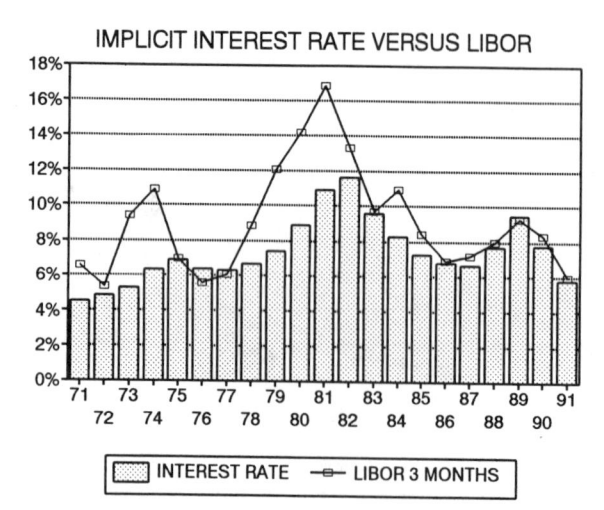

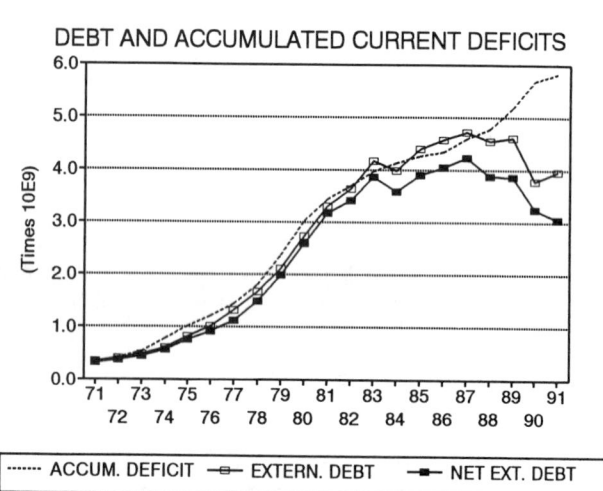

# COSTA RICA
## External Debt in US Dollars

# DOMINICAN REPUBLIC
## External Debt in US Dollars

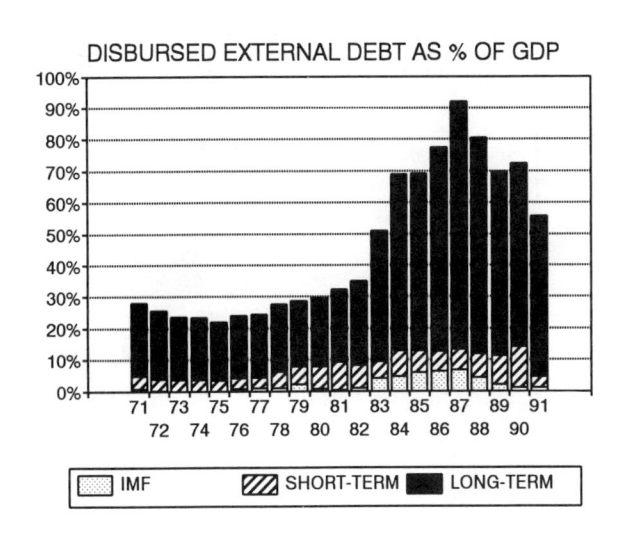

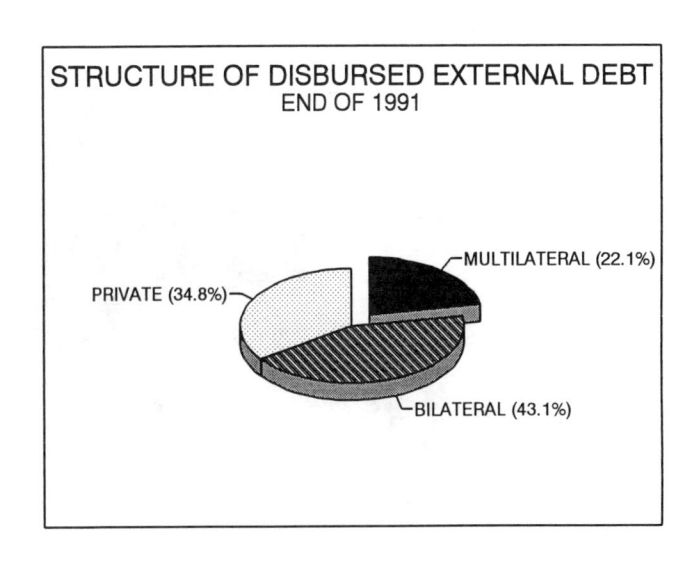

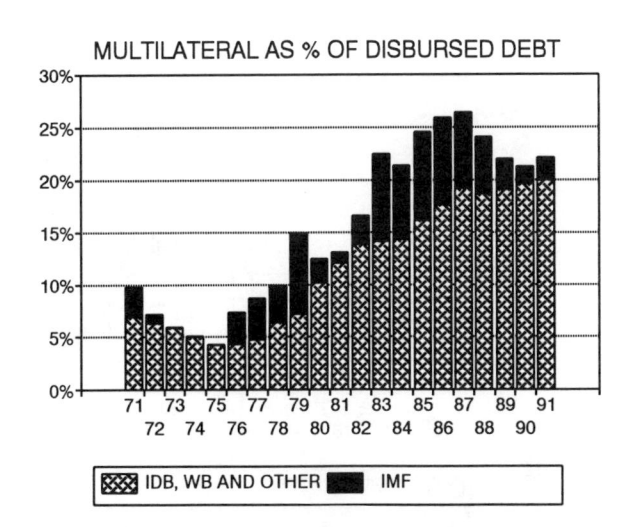

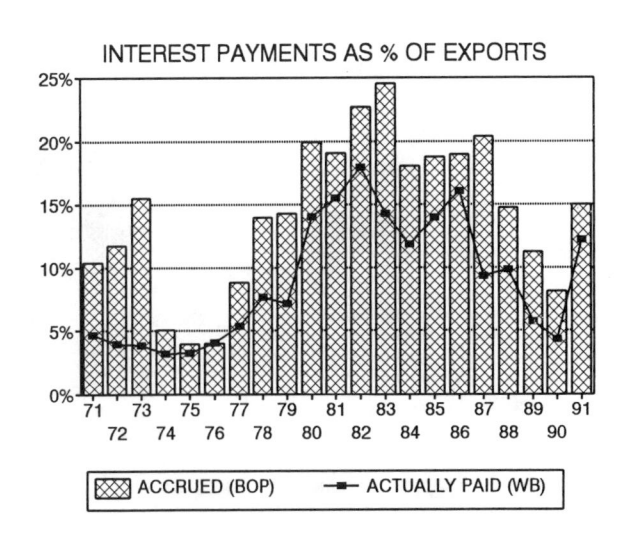

Statistics and Quantitative Analysis/IDB

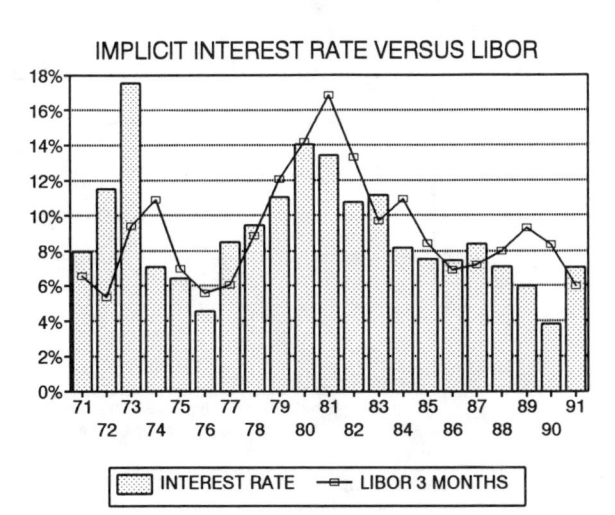

111

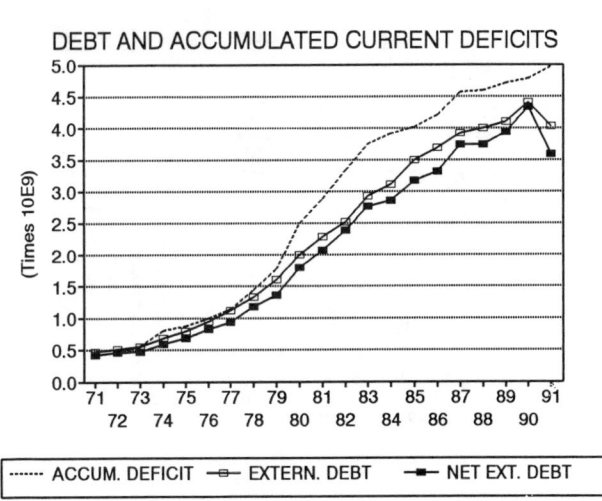

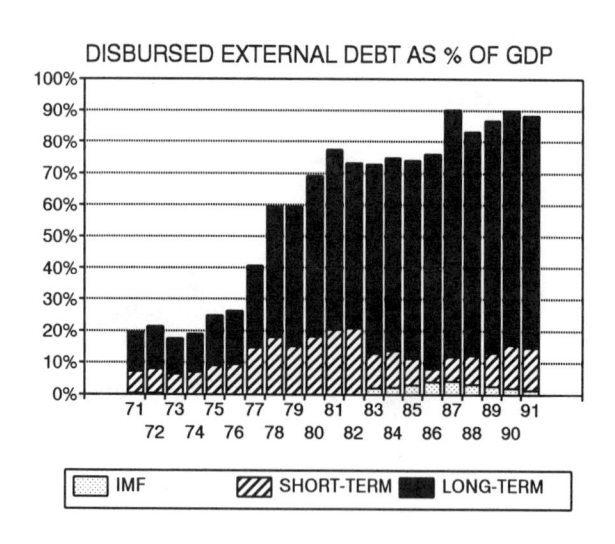

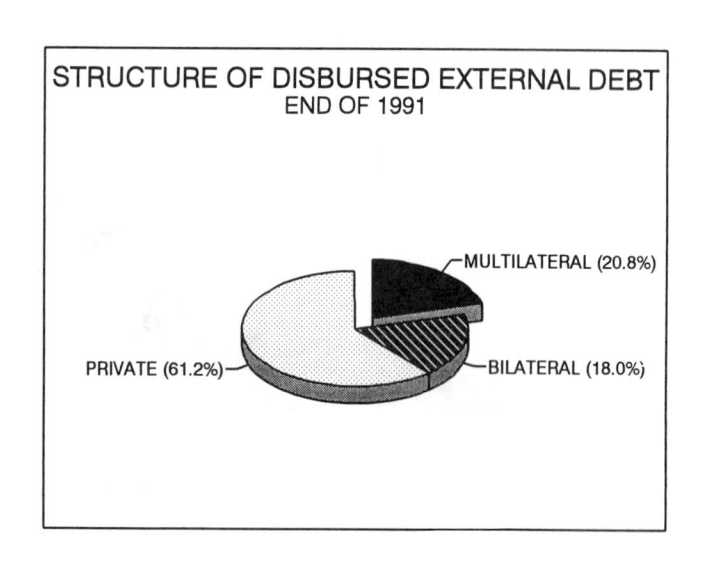

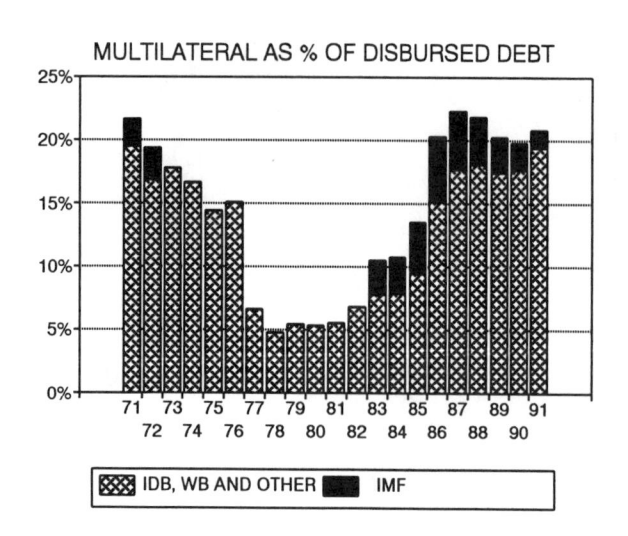

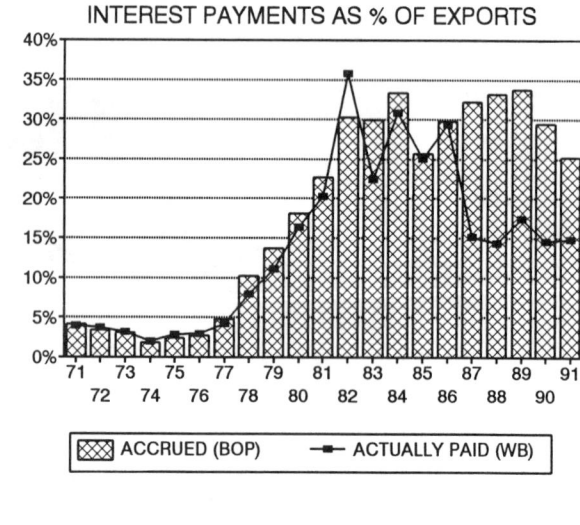

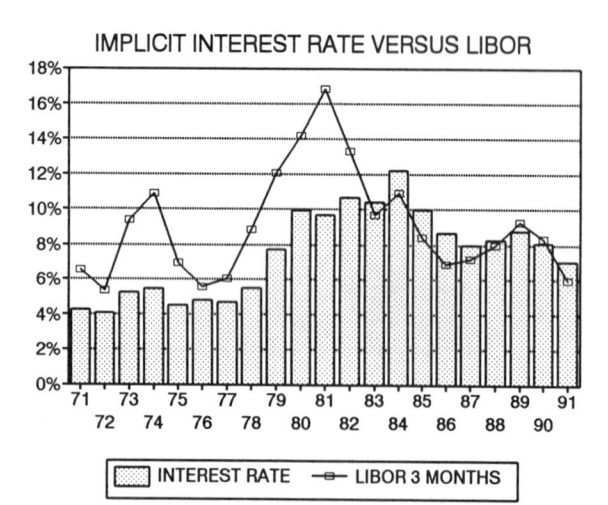

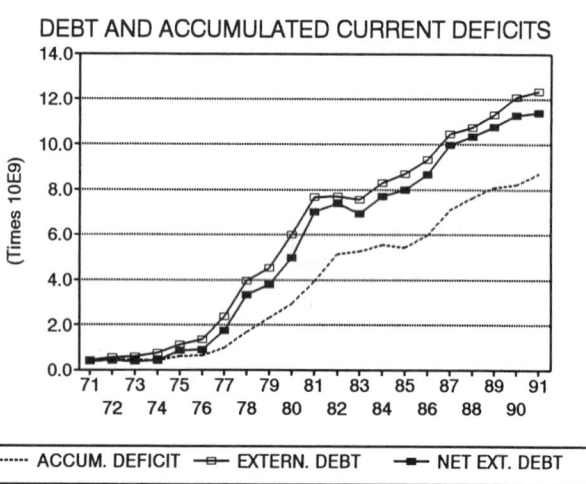

# ECUADOR
## External Debt in US Dollars

# EL SALVADOR
## External Debt in US Dollars

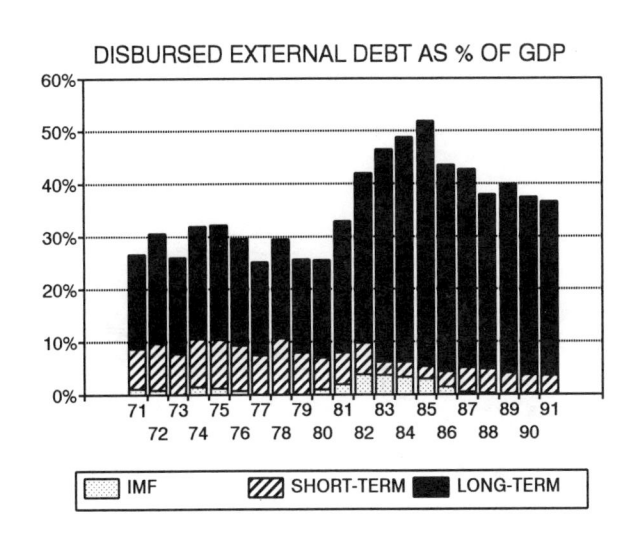

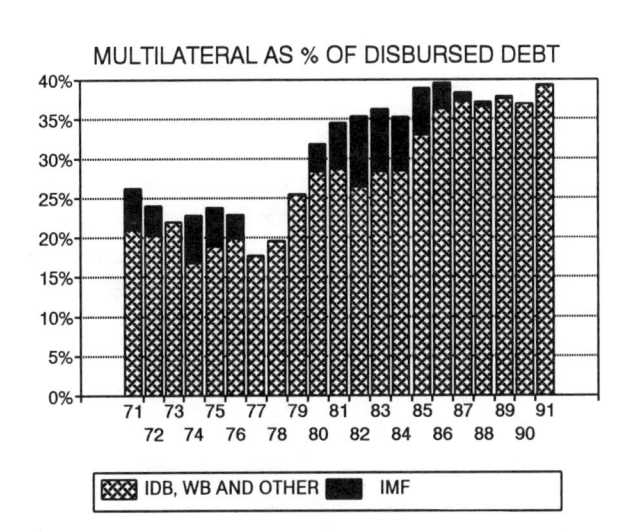

### DISBURSED EXTERNAL DEBT AS % OF GDP
IMF — SHORT-TERM — LONG-TERM

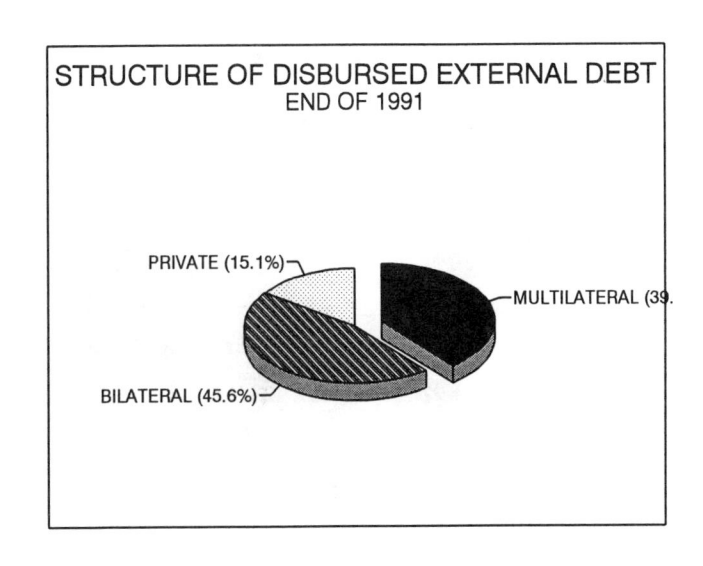

### STRUCTURE OF DISBURSED EXTERNAL DEBT
END OF 1991

- PRIVATE (15.1%)
- BILATERAL (45.6%)
- MULTILATERAL (39.)

### MULTILATERAL AS % OF DISBURSED DEBT
IDB, WB AND OTHER — IMF

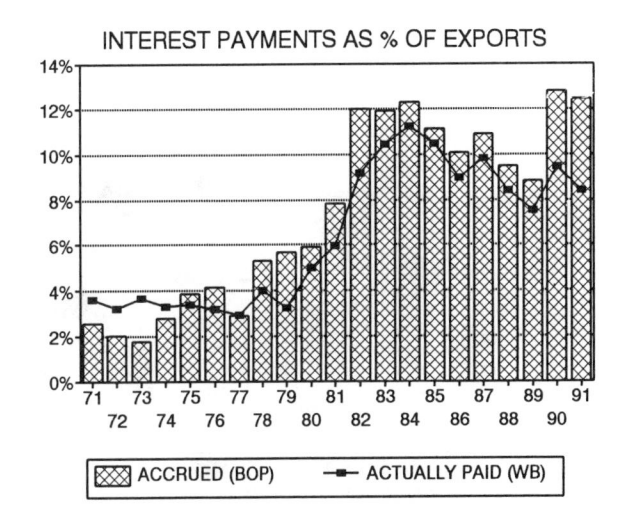

### INTEREST PAYMENTS AS % OF EXPORTS
ACCRUED (BOP) — ACTUALLY PAID (WB)

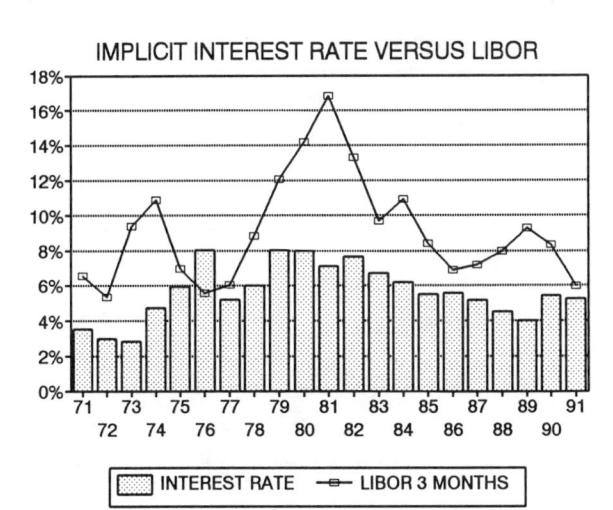

### IMPLICIT INTEREST RATE VERSUS LIBOR
INTEREST RATE — LIBOR 3 MONTHS

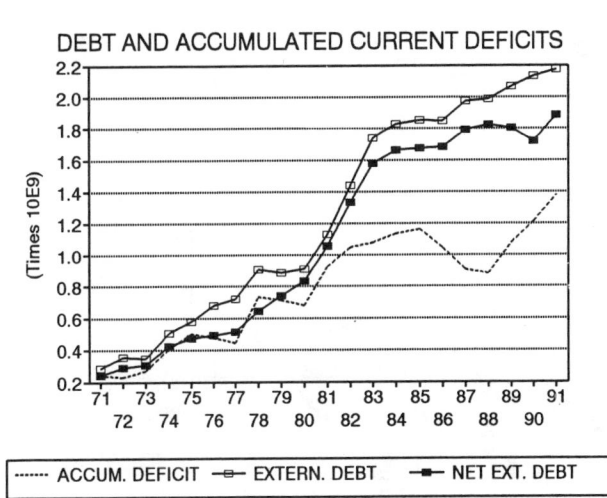

### DEBT AND ACCUMULATED CURRENT DEFICITS
(Times 10E9)
ACCUM. DEFICIT — EXTERN. DEBT — NET EXT. DEBT

Statistics and Quantitative Analysis/IDB

# GUATEMALA
## External Debt in US Dollars

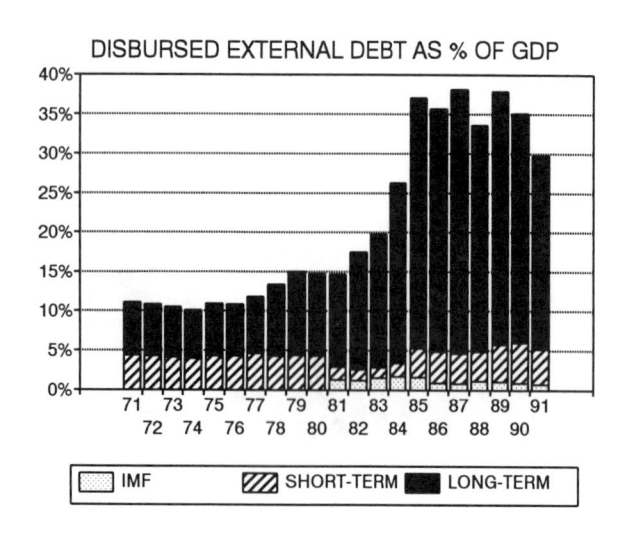

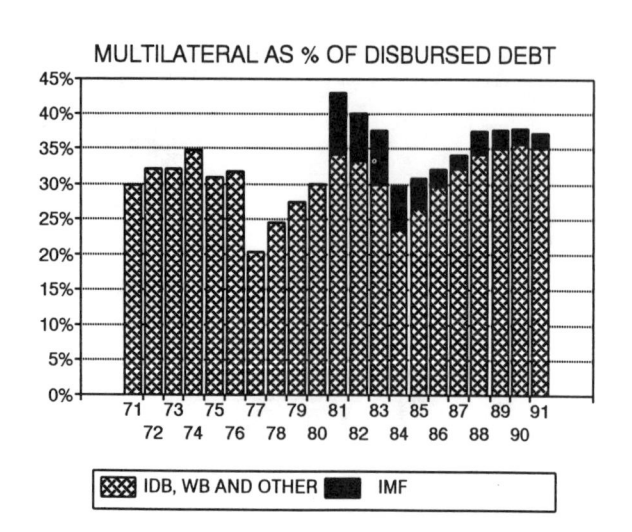

### DISBURSED EXTERNAL DEBT AS % OF GDP
IMF | SHORT-TERM | LONG-TERM

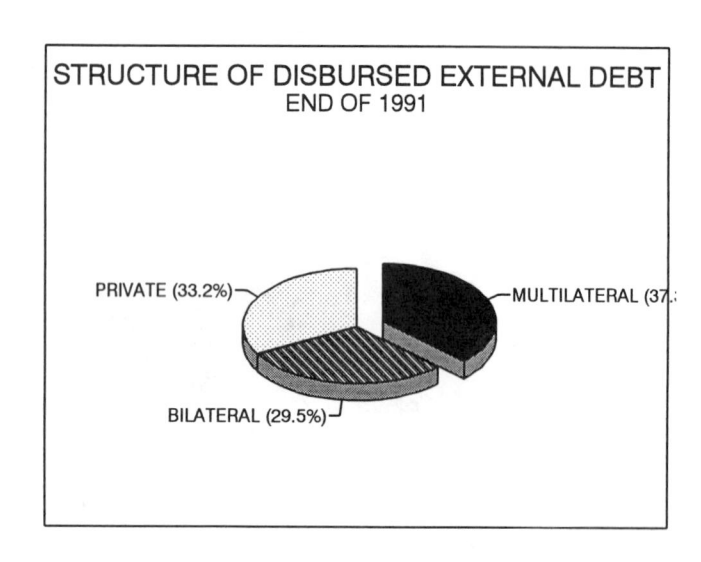

### STRUCTURE OF DISBURSED EXTERNAL DEBT
END OF 1991

- PRIVATE (33.2%)
- BILATERAL (29.5%)
- MULTILATERAL (37.;)

### MULTILATERAL AS % OF DISBURSED DEBT
IDB, WB AND OTHER | IMF

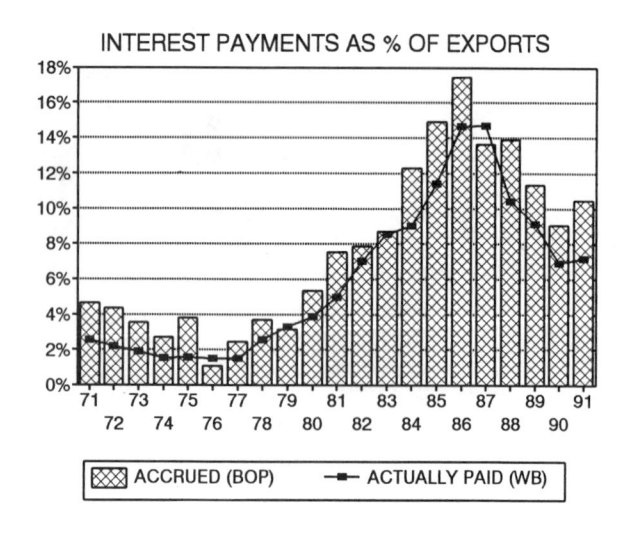

### INTEREST PAYMENTS AS % OF EXPORTS
ACCRUED (BOP) | ACTUALLY PAID (WB)

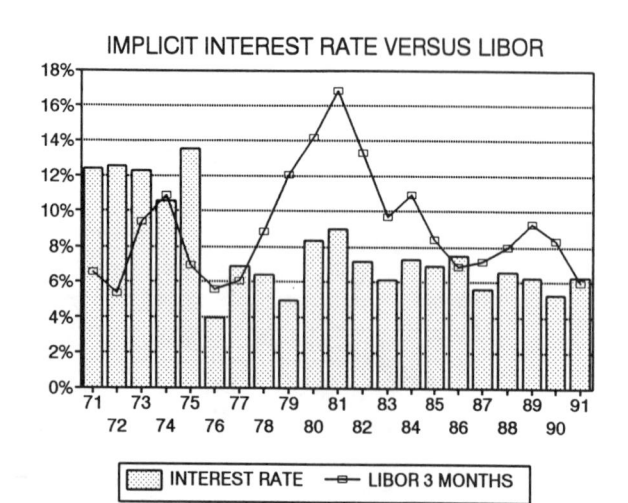

### IMPLICIT INTEREST RATE VERSUS LIBOR
INTEREST RATE | LIBOR 3 MONTHS

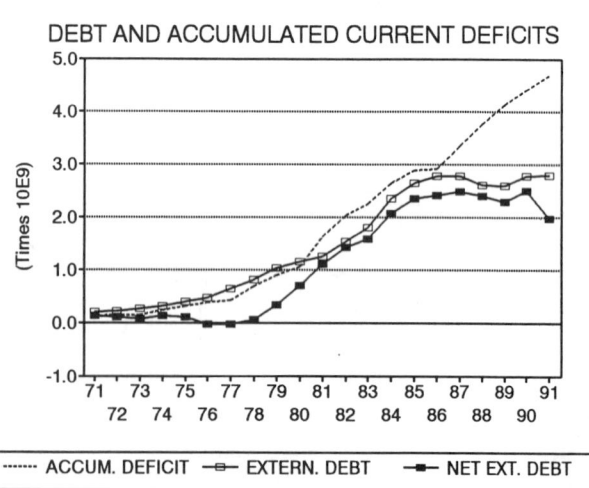

### DEBT AND ACCUMULATED CURRENT DEFICITS
(Times 10E9)
ACCUM. DEFICIT | EXTERN. DEBT | NET EXT. DEBT

Statistics and Quantitative Analysis/IDB

# GUYANA
## External Debt in US Dollars

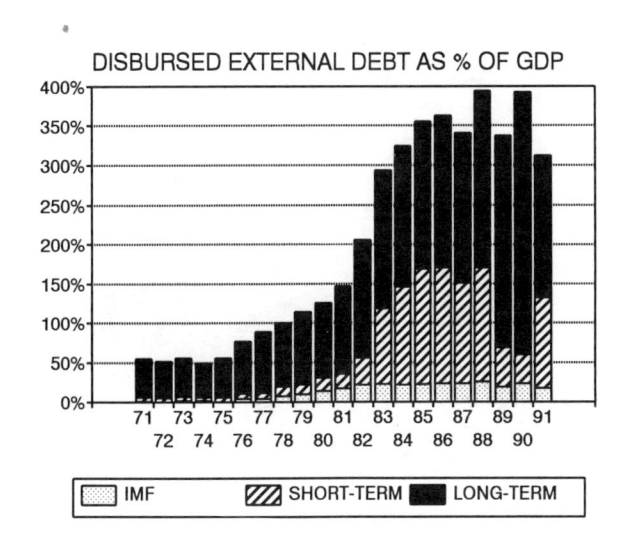

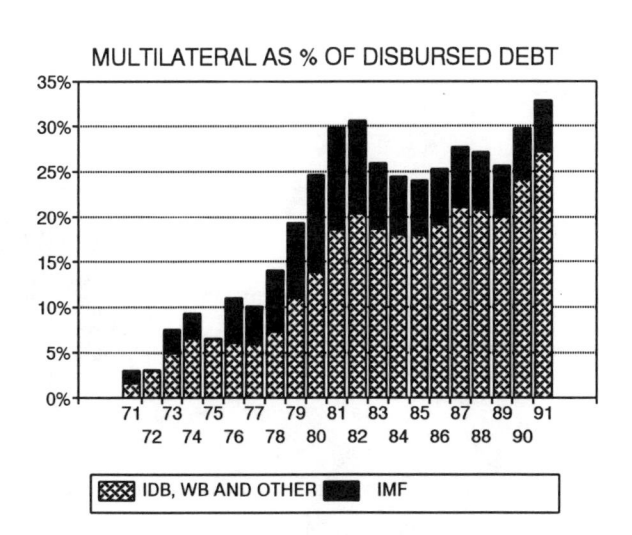

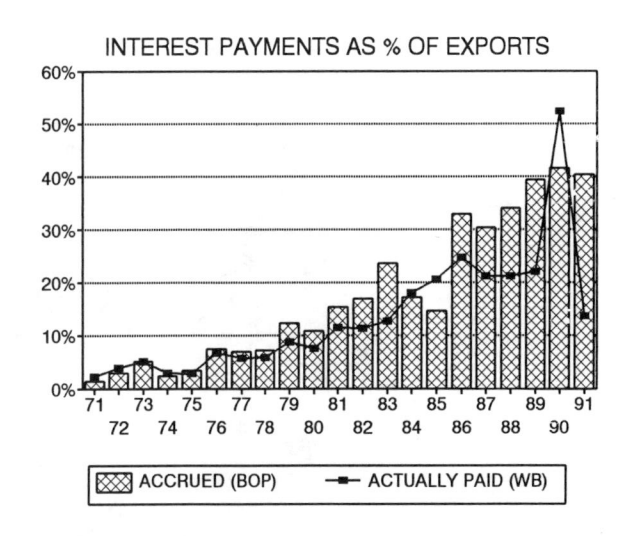

### DISBURSED EXTERNAL DEBT AS % OF GDP
IMF — SHORT-TERM — LONG-TERM

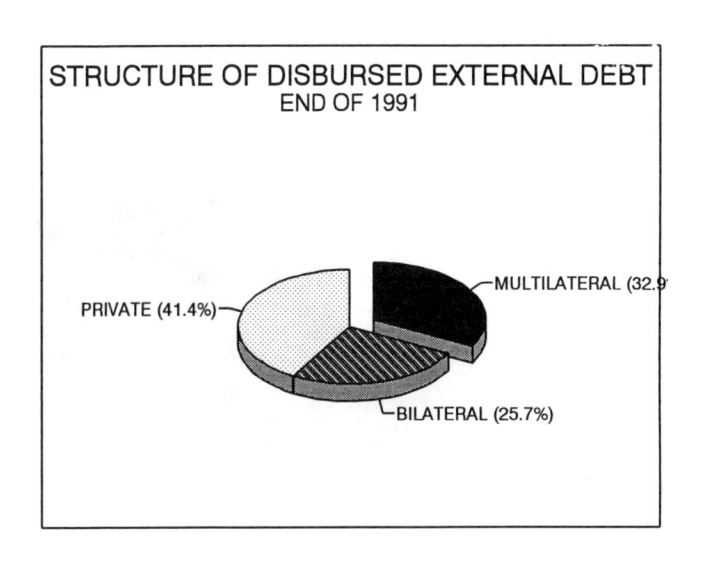

### STRUCTURE OF DISBURSED EXTERNAL DEBT
END OF 1991

- MULTILATERAL (32.9%)
- BILATERAL (25.7%)
- PRIVATE (41.4%)

### MULTILATERAL AS % OF DISBURSED DEBT
IDB, WB AND OTHER — IMF

### INTEREST PAYMENTS AS % OF EXPORTS
ACCRUED (BOP) — ACTUALLY PAID (WB)

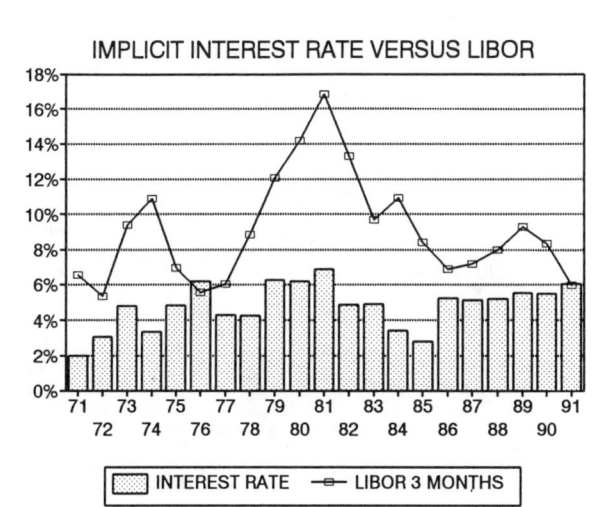

### IMPLICIT INTEREST RATE VERSUS LIBOR
INTEREST RATE — LIBOR 3 MONTHS

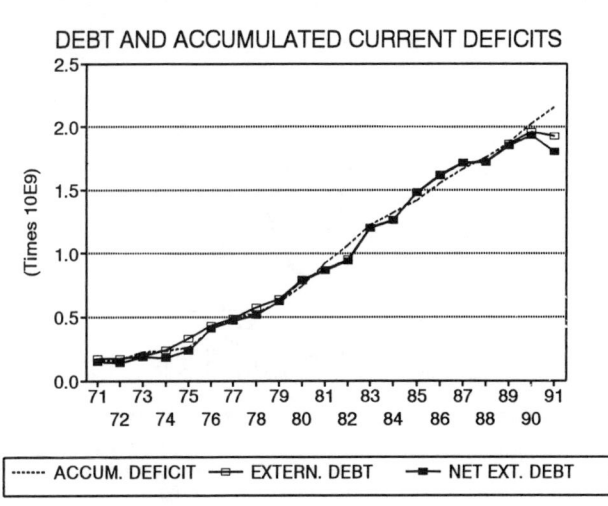

### DEBT AND ACCUMULATED CURRENT DEFICITS
(Times 10E9)
ACCUM. DEFICIT — EXTERN. DEBT — NET EXT. DEBT

Statistics and Quantitative Analysis/IDB

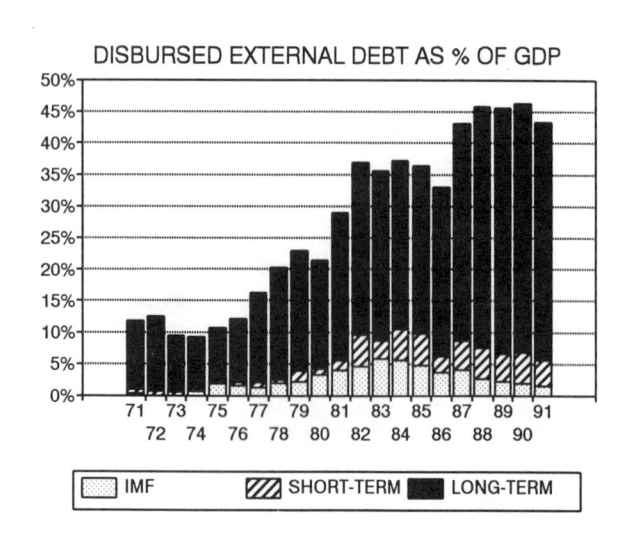

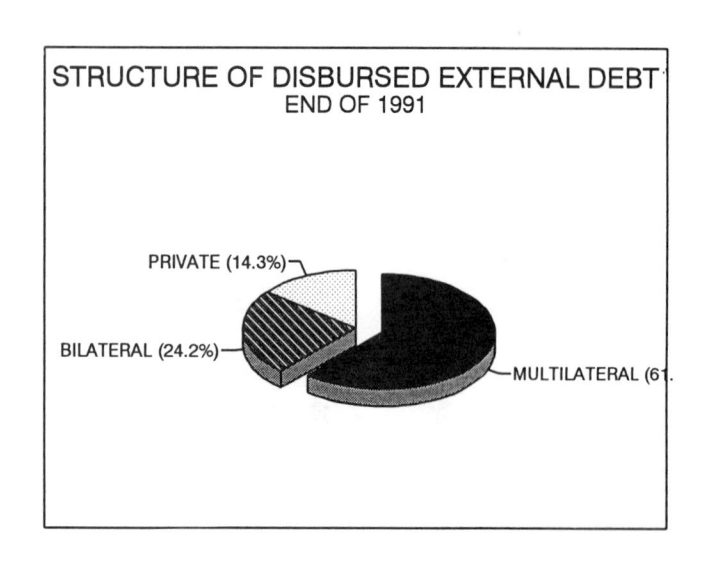

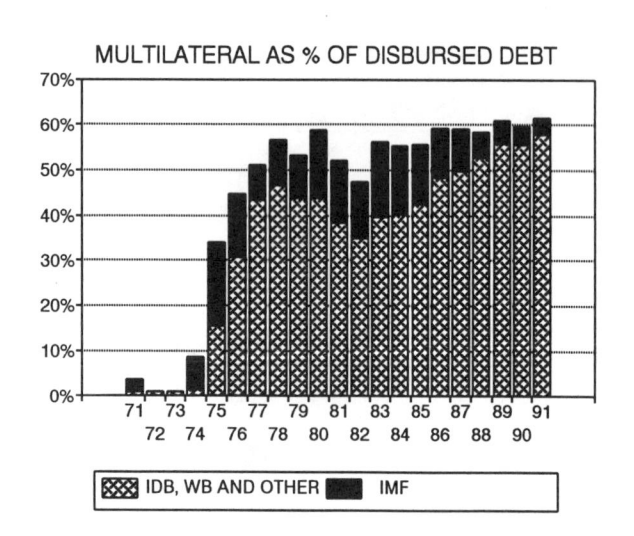

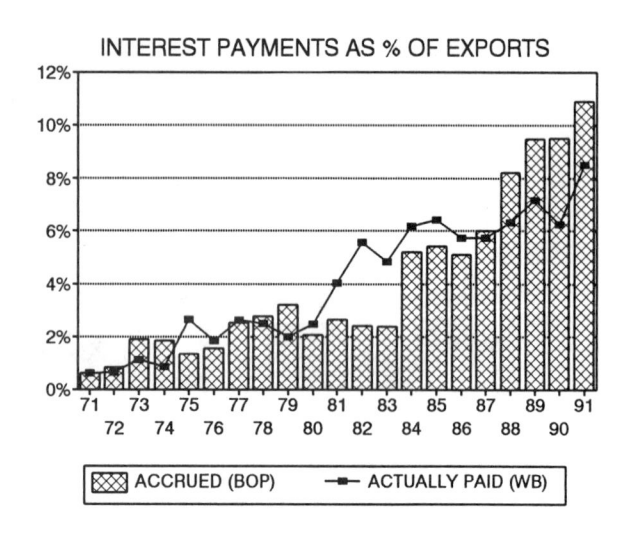

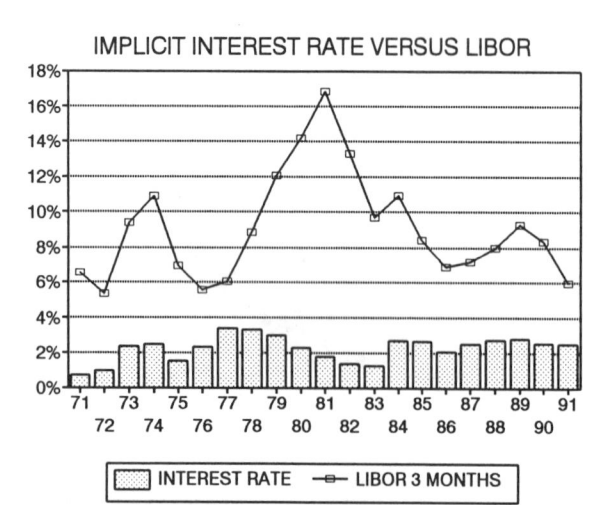

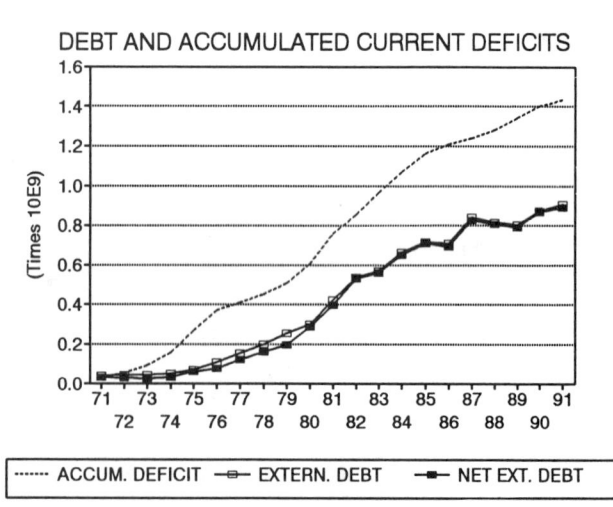

# HAITI
## External Debt in US Dollars

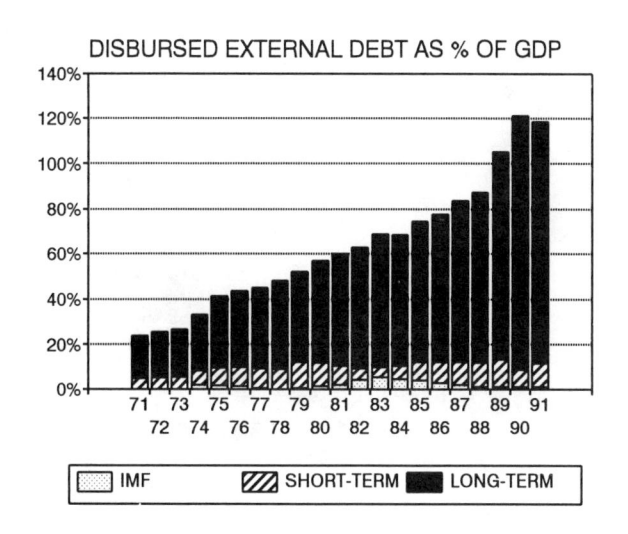

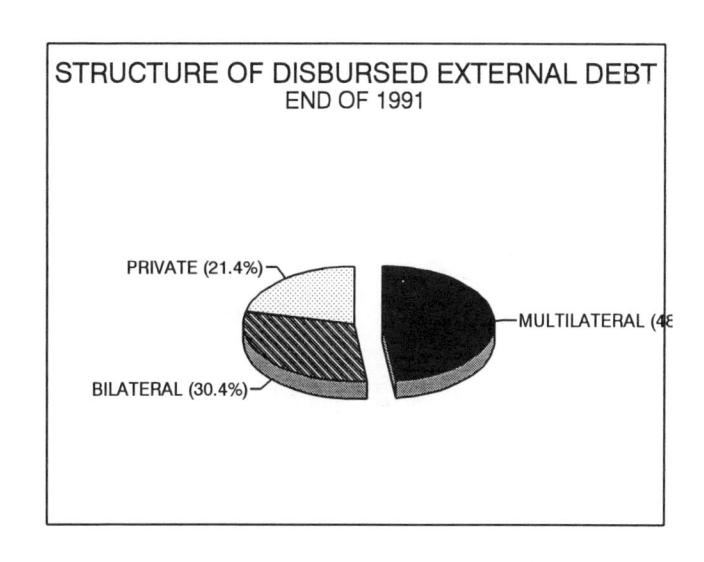

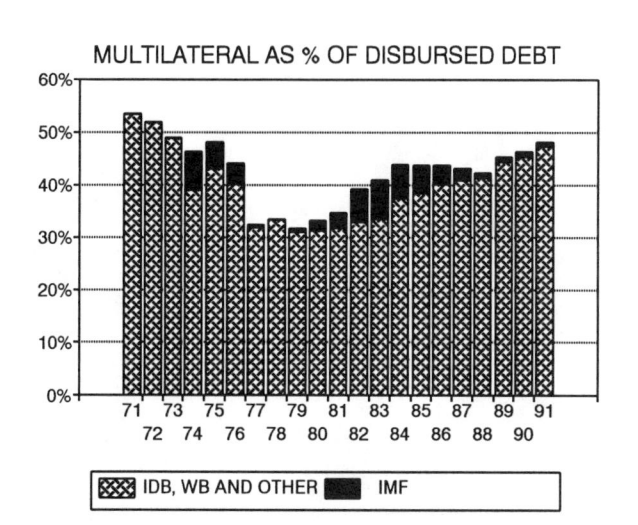

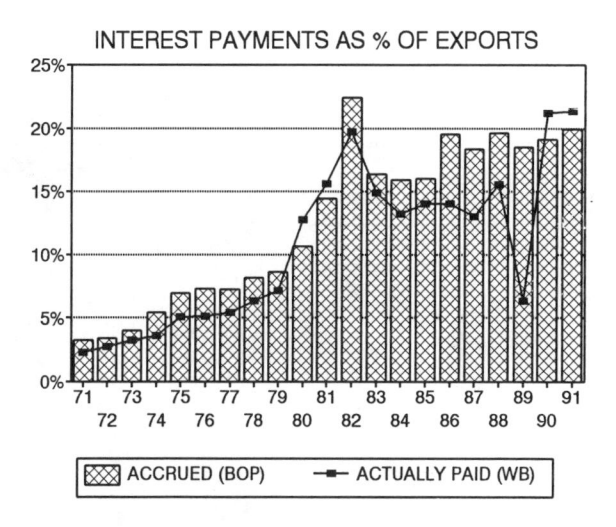

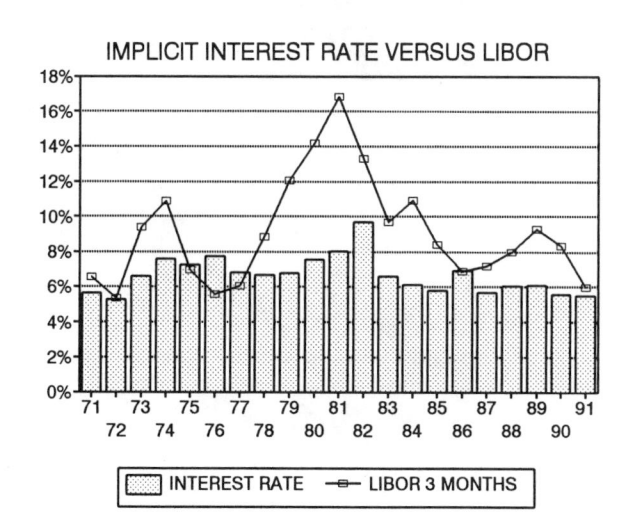

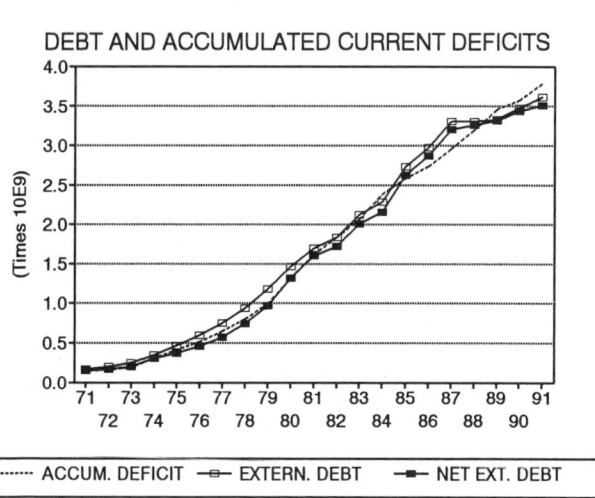

# HONDURAS
## External Debt in US Dollars

Statistics and Quantitative Analysis/IDB

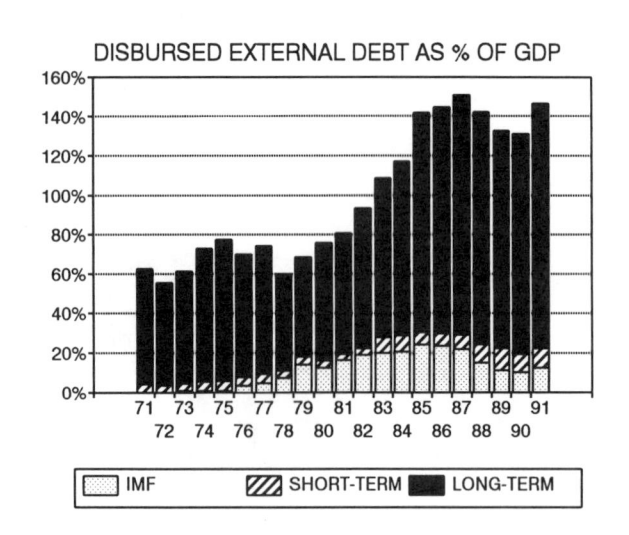

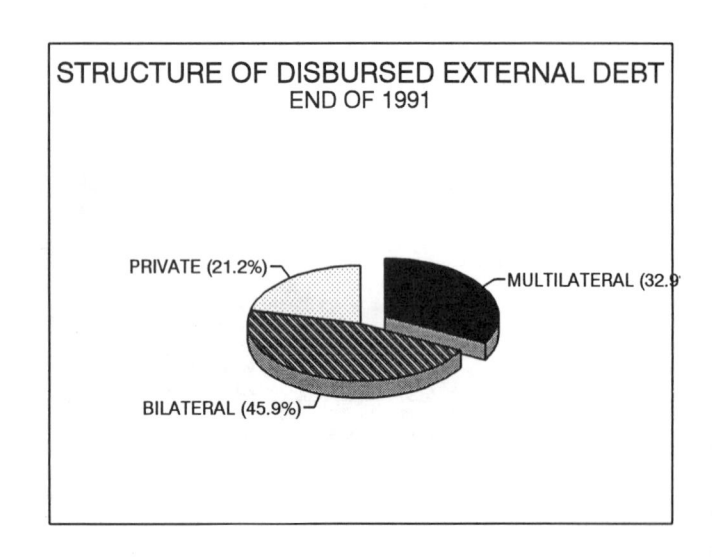

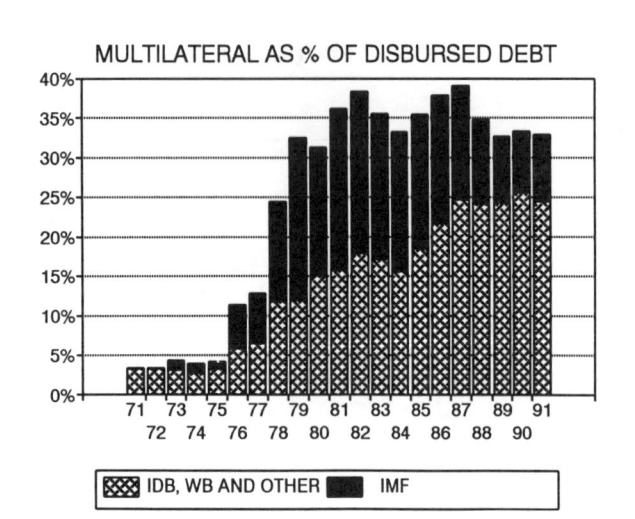

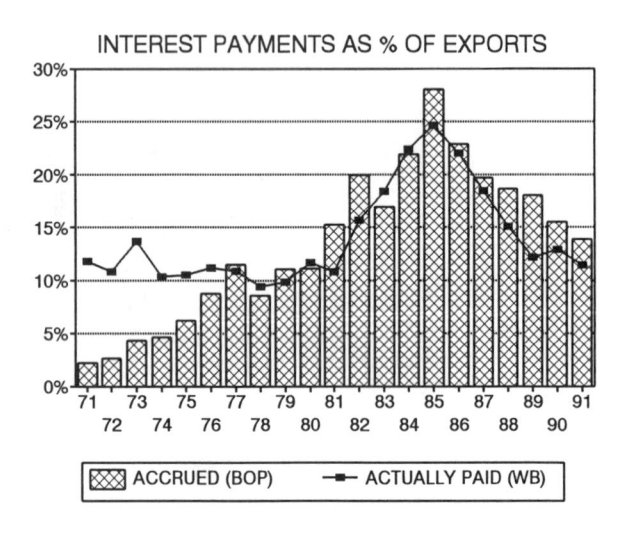

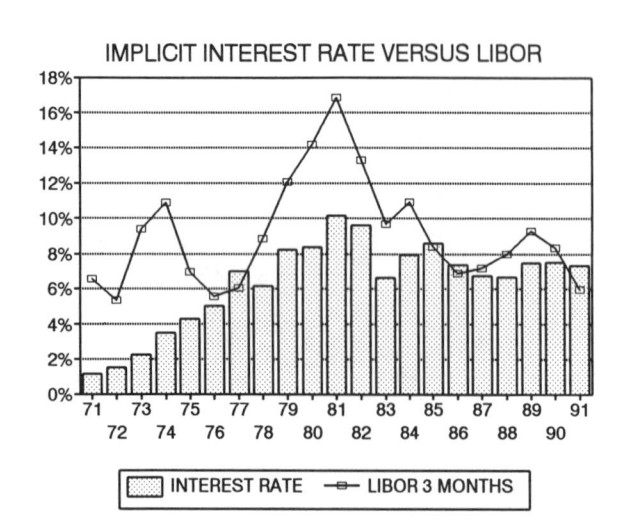

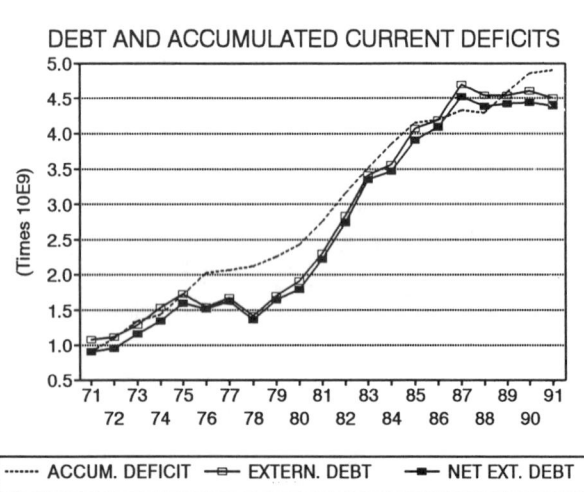

# JAMAICA
## External Debt in US Dollars

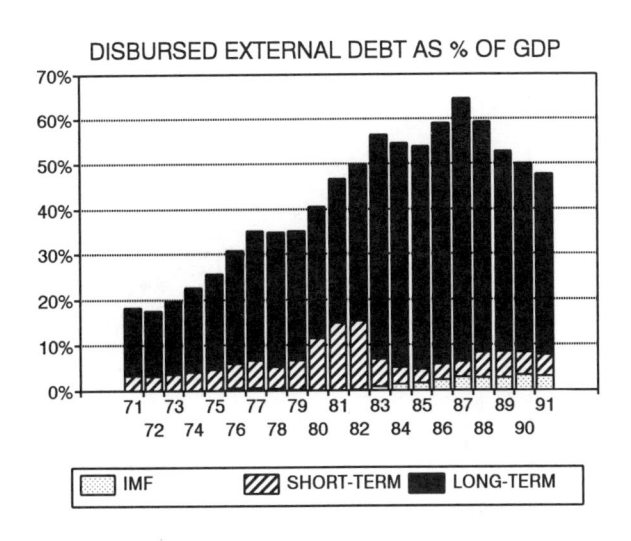

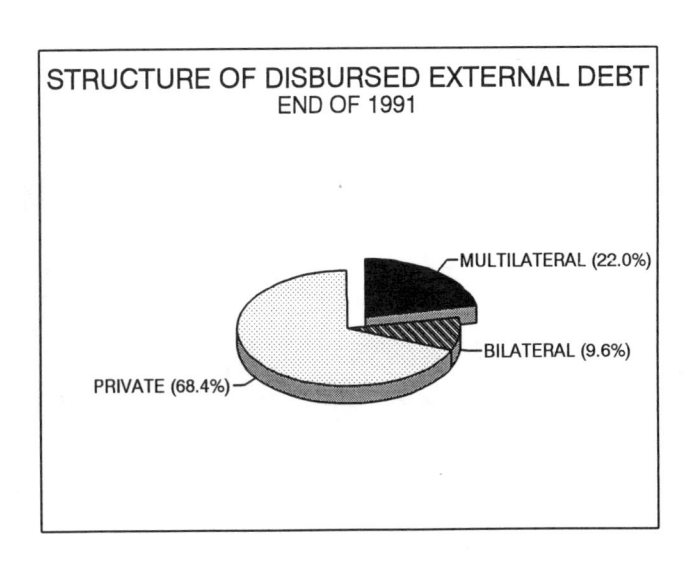

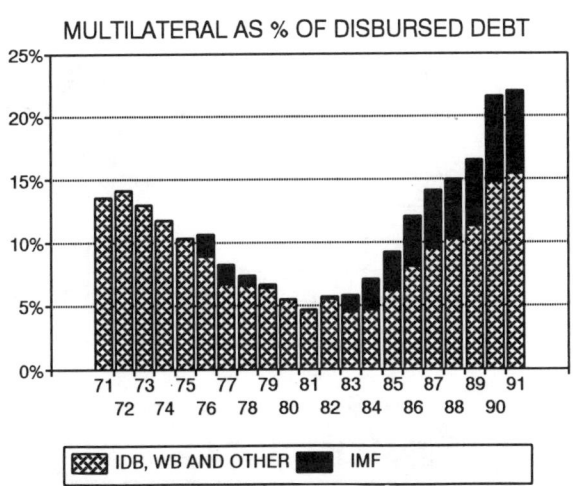

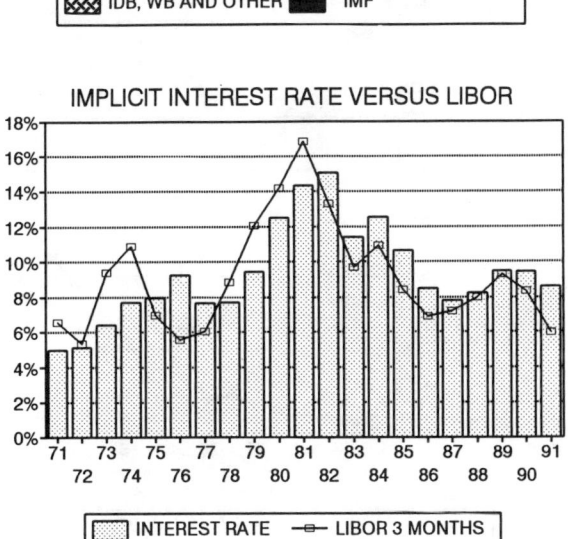

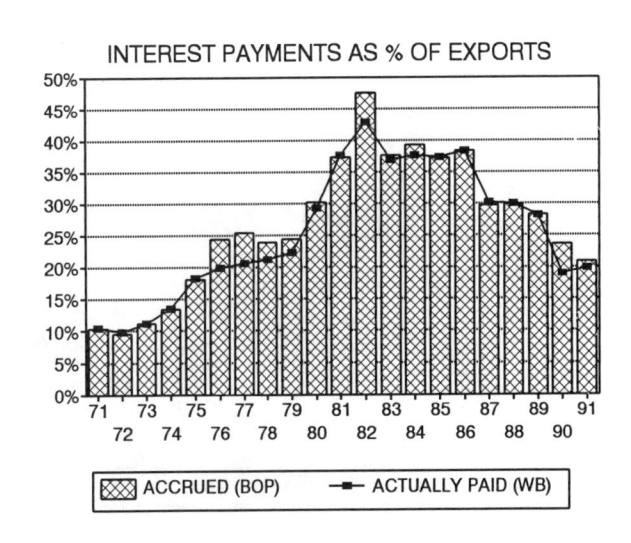

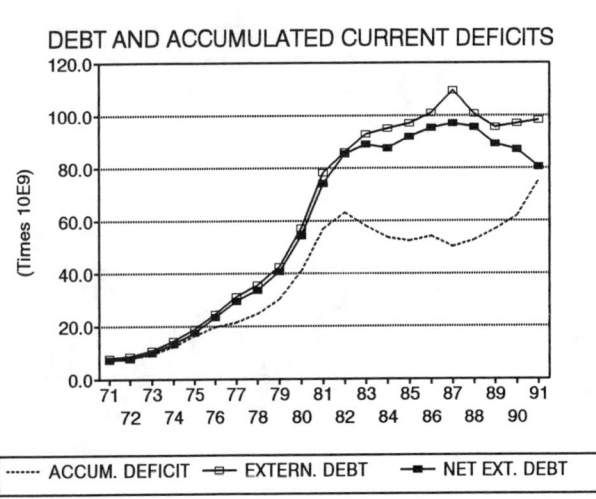

# MEXICO
## External Debt in US Dollars

Statistics and Quantitative Analysis/IDB

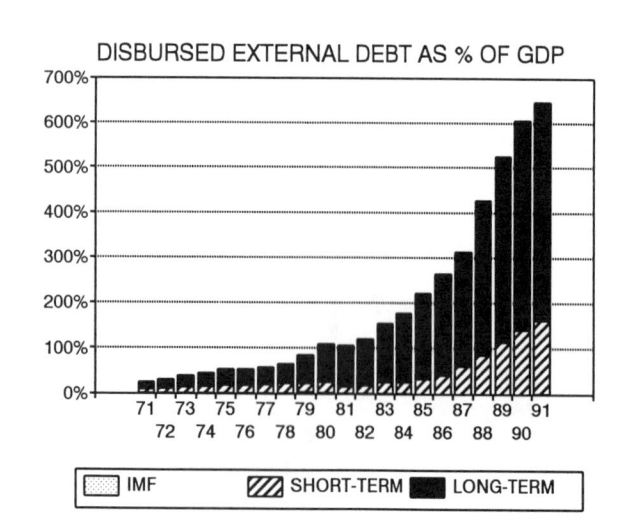

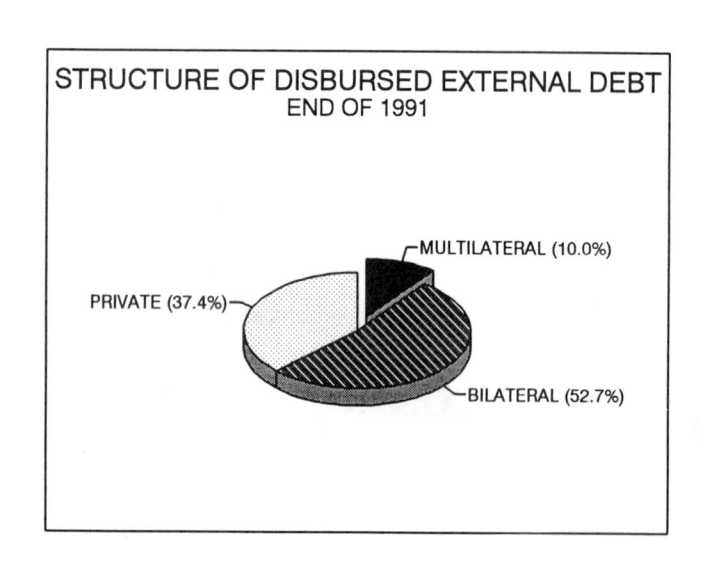

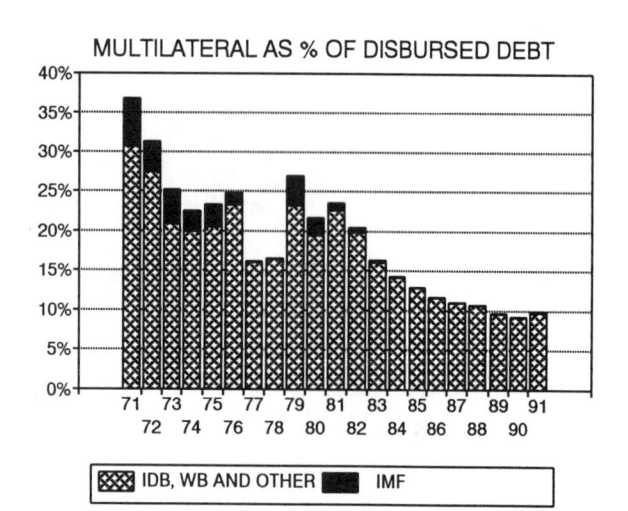

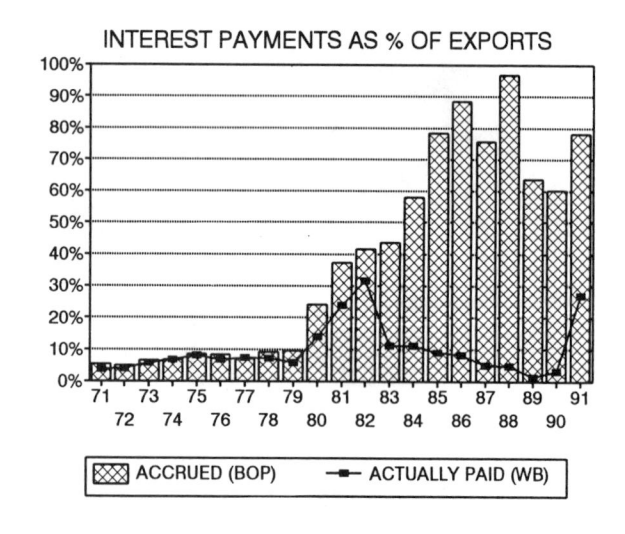

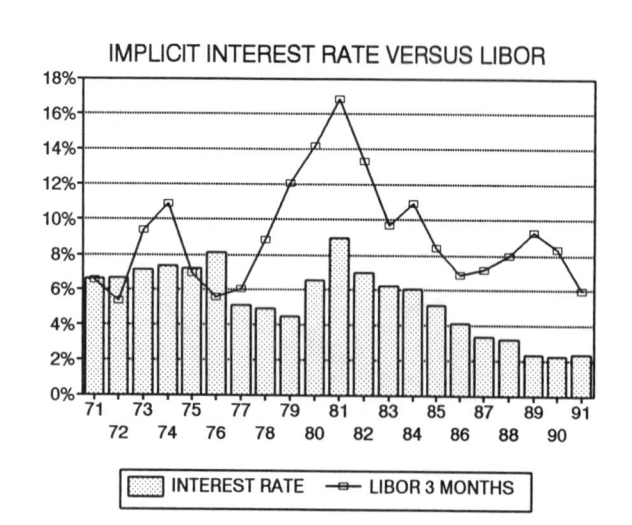

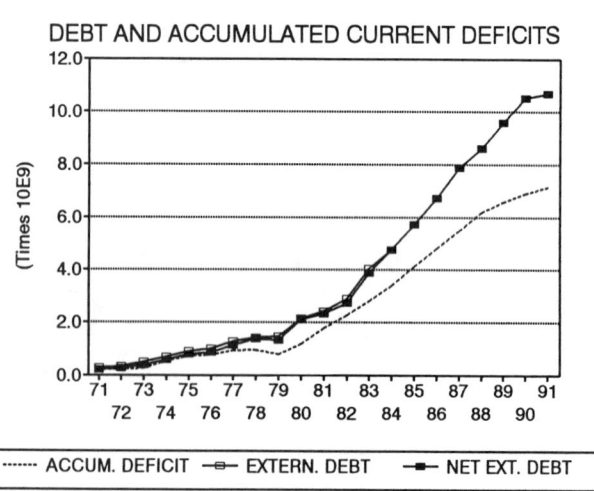

# NICARAGUA
## External Debt in US Dollars

Statistics and Quantitative Analysis/IDB

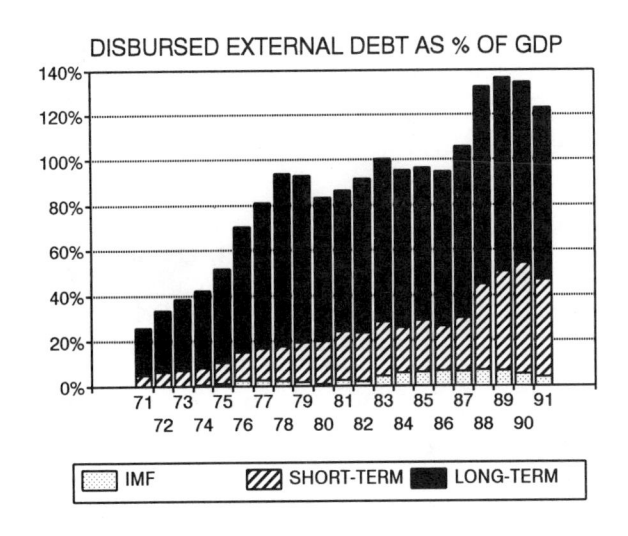

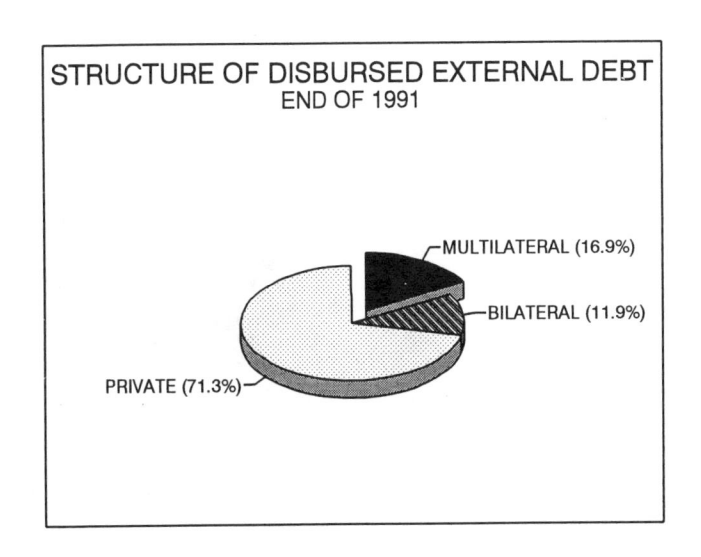

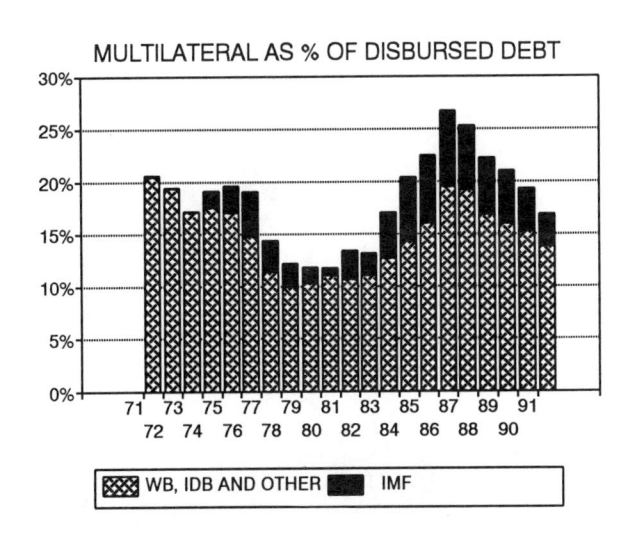

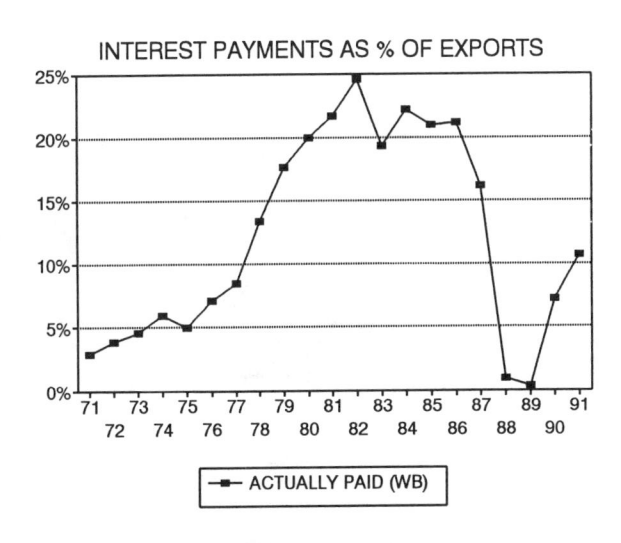

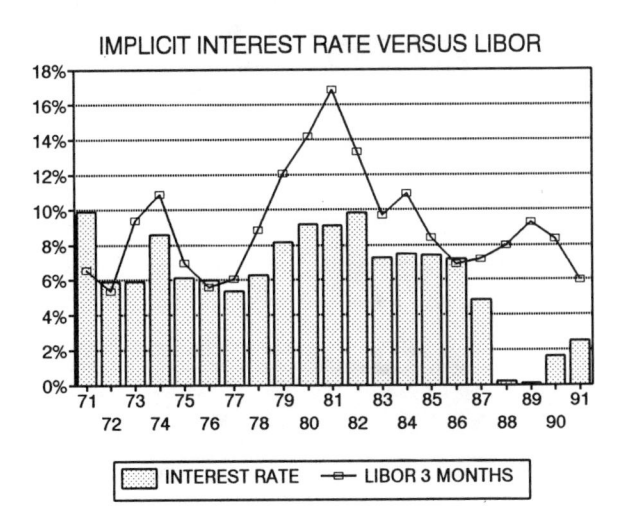

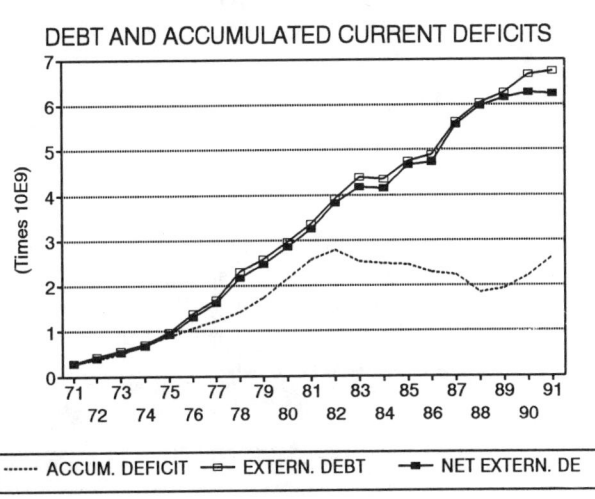

# PANAMA
## External Debt in US Dollars

Statistics and Quantitative Analysis/IDB

# PARAGUAY
## External Debt in US Dollars

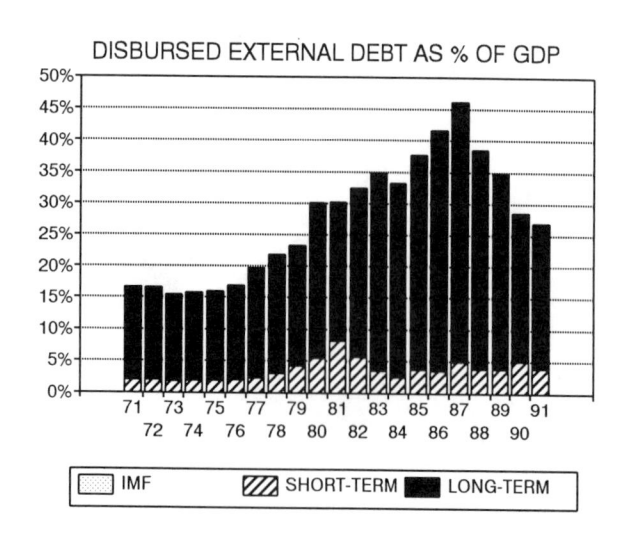

### DISBURSED EXTERNAL DEBT AS % OF GDP
IMF — SHORT-TERM — LONG-TERM

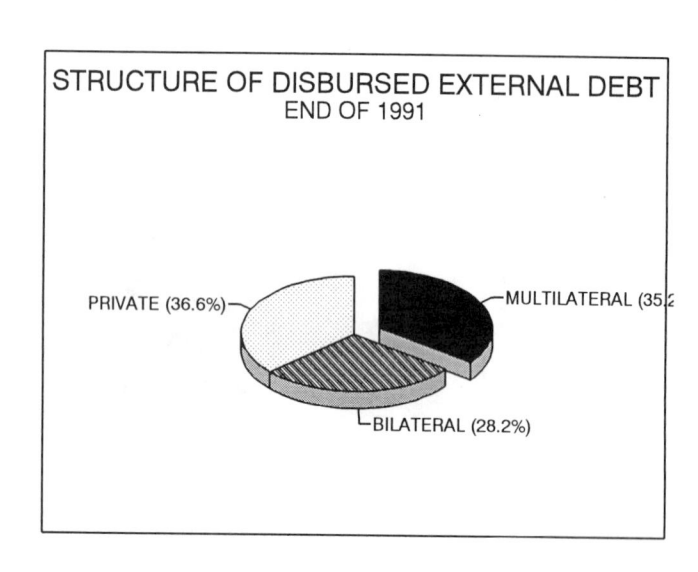

### STRUCTURE OF DISBURSED EXTERNAL DEBT
END OF 1991

- PRIVATE (36.6%)
- MULTILATERAL (35.2%)
- BILATERAL (28.2%)

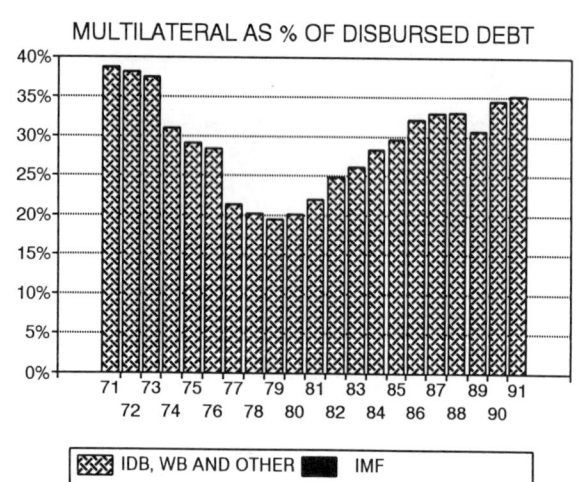

### MULTILATERAL AS % OF DISBURSED DEBT
IDB, WB AND OTHER — IMF

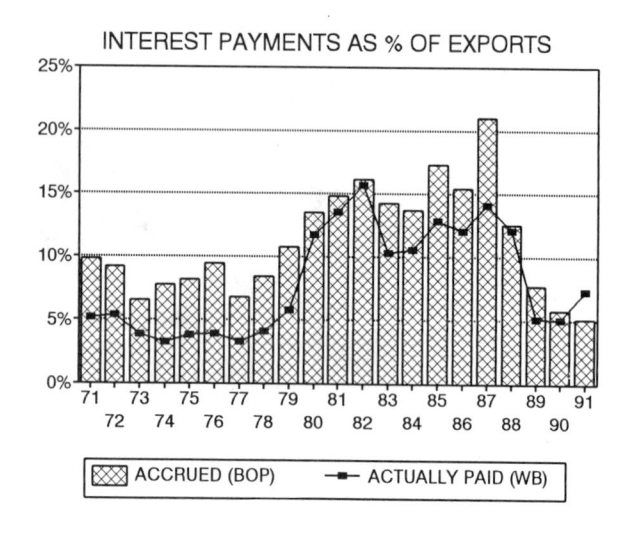

### INTEREST PAYMENTS AS % OF EXPORTS
ACCRUED (BOP) — ACTUALLY PAID (WB)

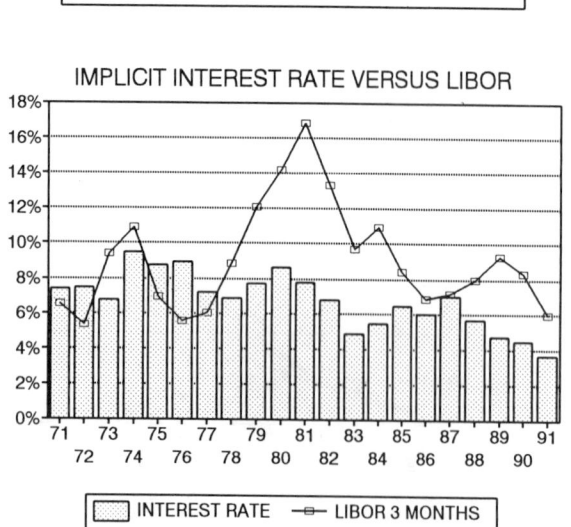

### IMPLICIT INTEREST RATE VERSUS LIBOR
INTEREST RATE — LIBOR 3 MONTHS

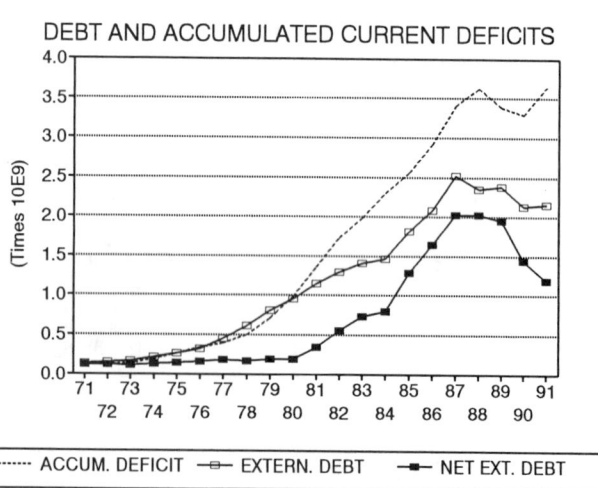

### DEBT AND ACCUMULATED CURRENT DEFICITS
(Times 10E9)
ACCUM. DEFICIT — EXTERN. DEBT — NET EXT. DEBT

Statistics and Quantitative Analysis/IDB

# PERU
## External Debt in US Dollars

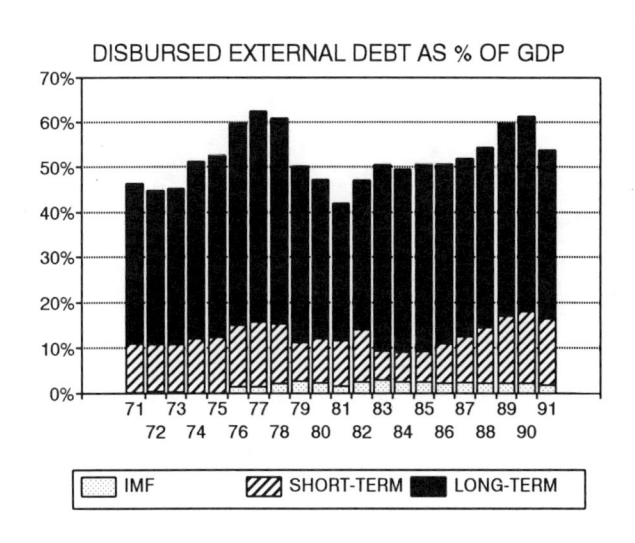

### DISBURSED EXTERNAL DEBT AS % OF GDP
(IMF, SHORT-TERM, LONG-TERM)

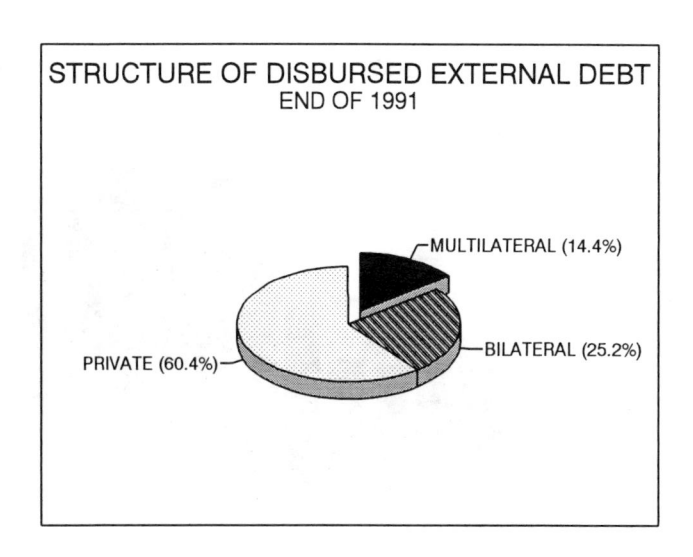

### STRUCTURE OF DISBURSED EXTERNAL DEBT
END OF 1991

- MULTILATERAL (14.4%)
- BILATERAL (25.2%)
- PRIVATE (60.4%)

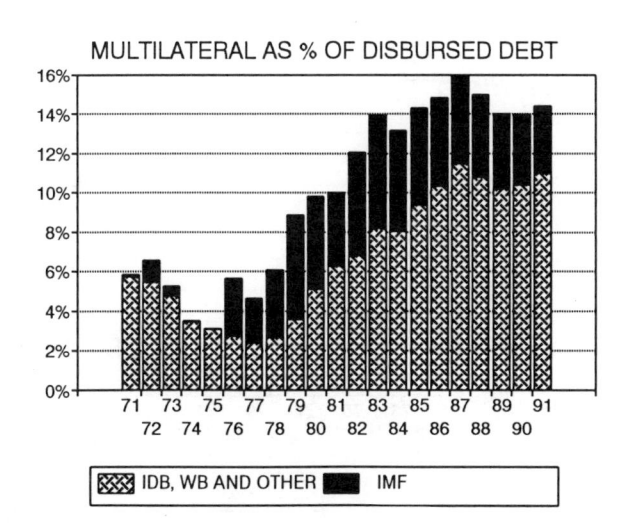

### MULTILATERAL AS % OF DISBURSED DEBT
(IDB, WB AND OTHER; IMF)

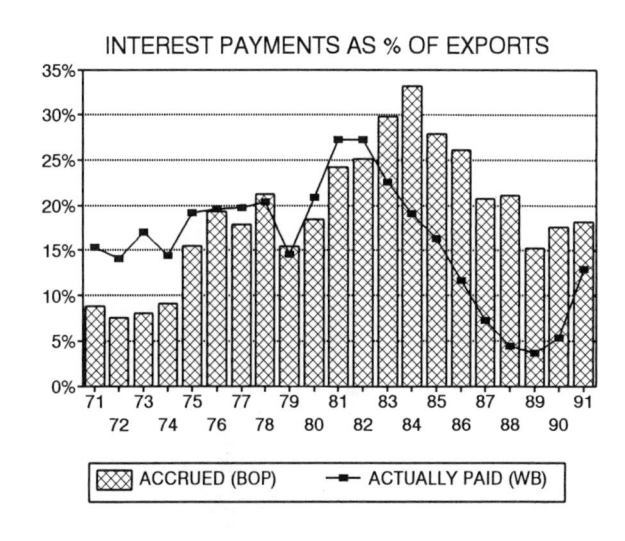

### INTEREST PAYMENTS AS % OF EXPORTS
(ACCRUED (BOP); ACTUALLY PAID (WB))

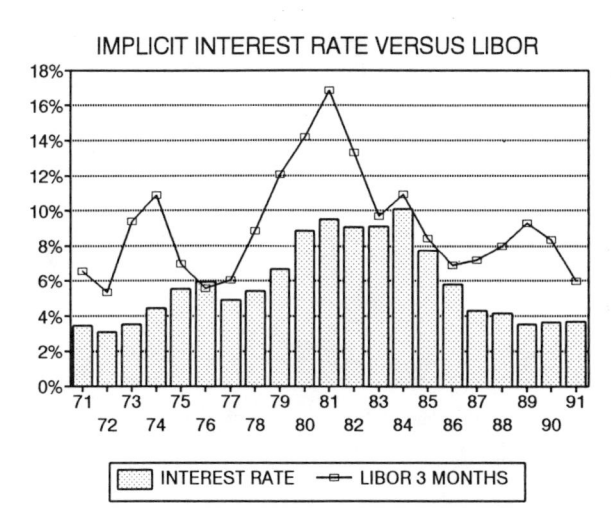

### IMPLICIT INTEREST RATE VERSUS LIBOR
(INTEREST RATE; LIBOR 3 MONTHS)

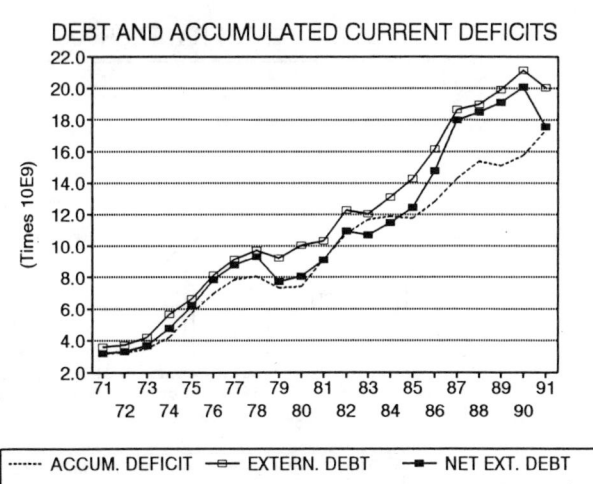

### DEBT AND ACCUMULATED CURRENT DEFICITS
(Times 10E9)
(ACCUM. DEFICIT; EXTERN. DEBT; NET EXT. DEBT)

Statistics and Quantitative Analysis/IDB

# SURINAME
## External Debt in US Dollars

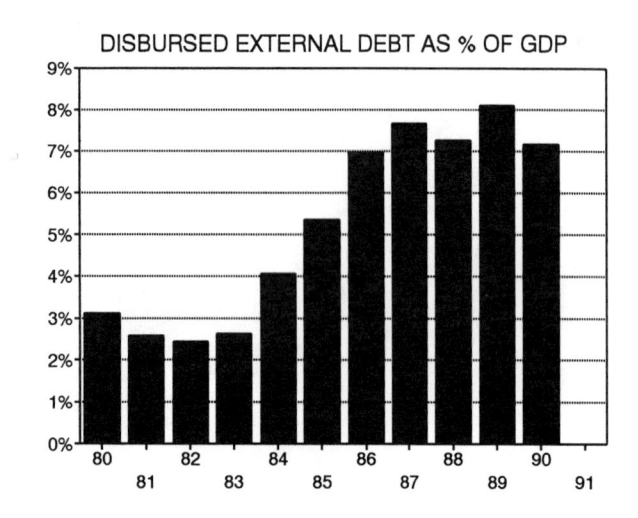

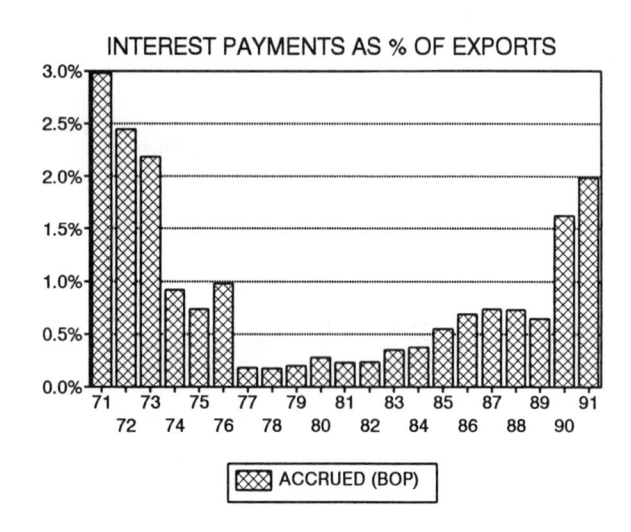

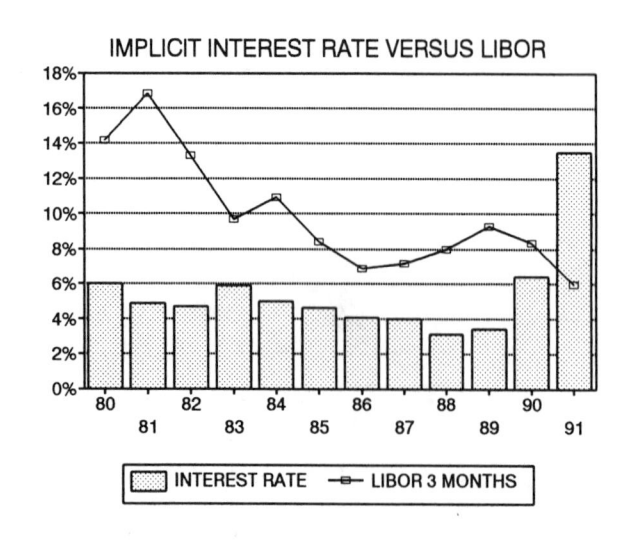

DISBURSED EXTERNAL DEBT AS % OF GDP

INTEREST PAYMENTS AS % OF EXPORTS

ACCRUED (BOP)

IMPLICIT INTEREST RATE VERSUS LIBOR

INTEREST RATE — LIBOR 3 MONTHS

Statistics and Quantitative Analysis/IDB

# TRINIDAD AND TOBAGO
## External Debt in US Dollars

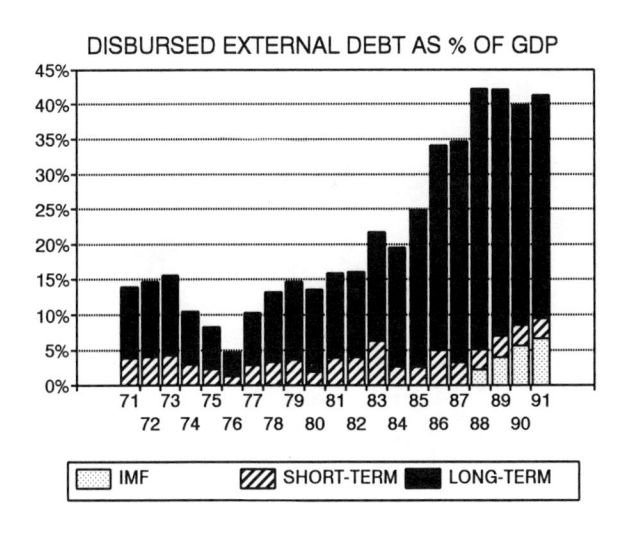

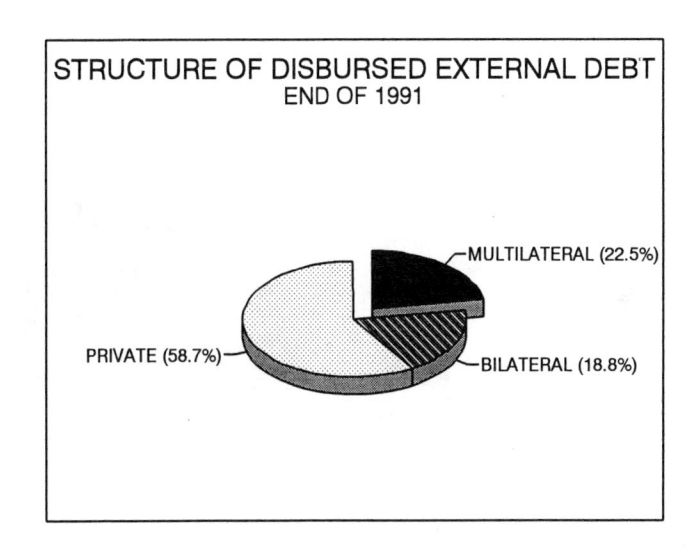

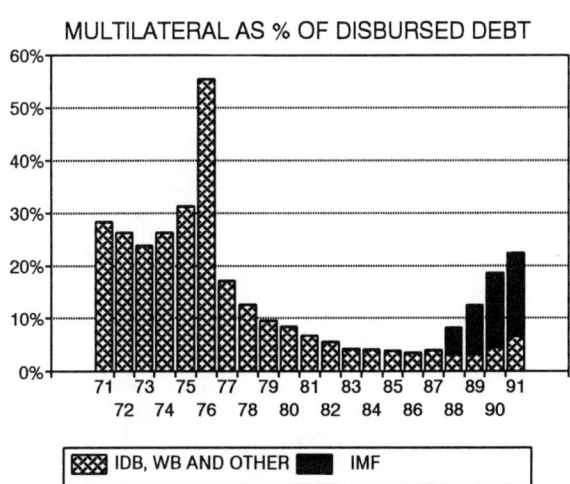

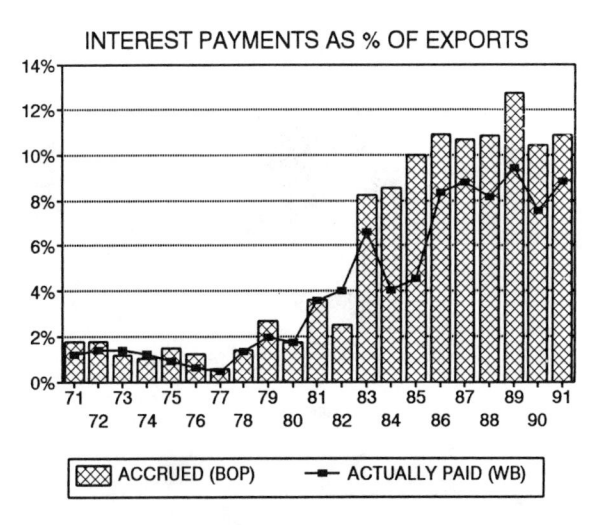

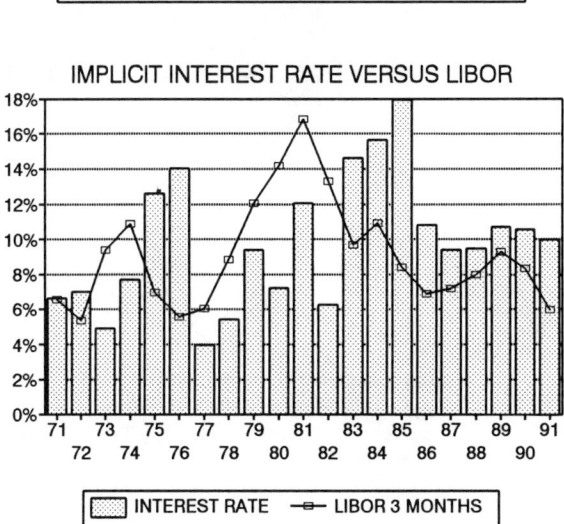

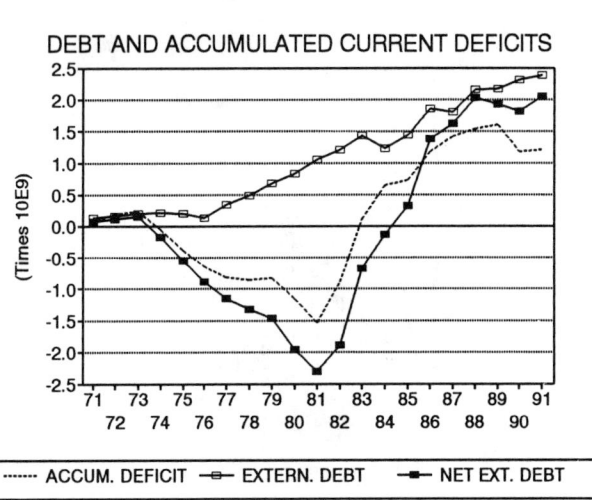

Statistics and Quantitative Analysis/IDB

125

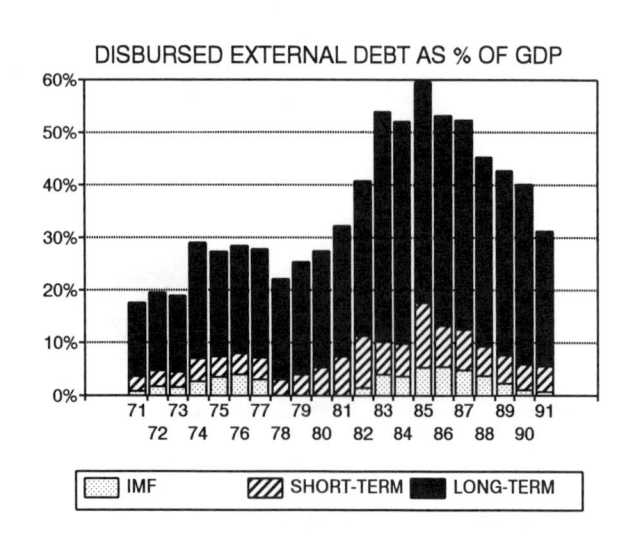

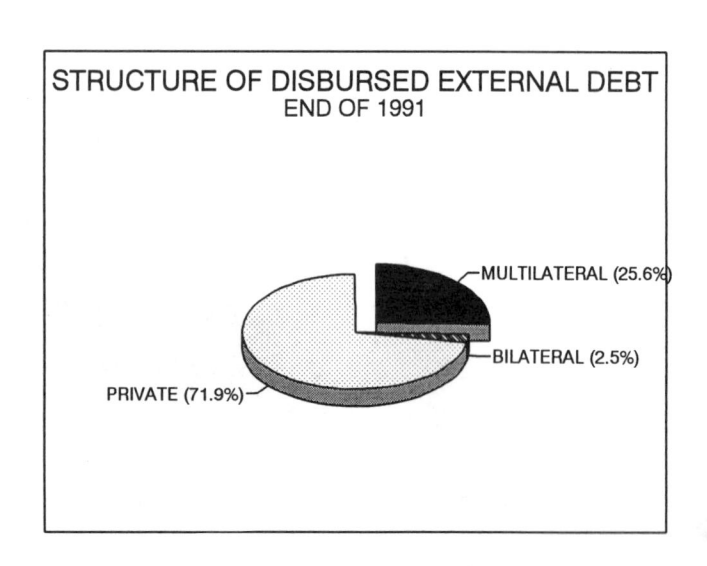

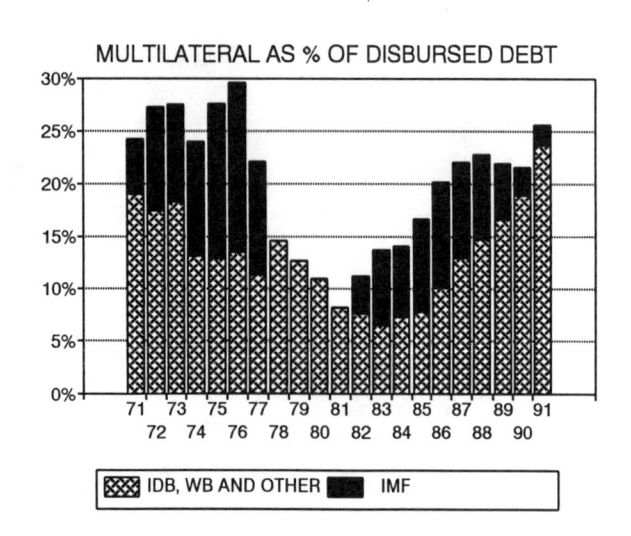

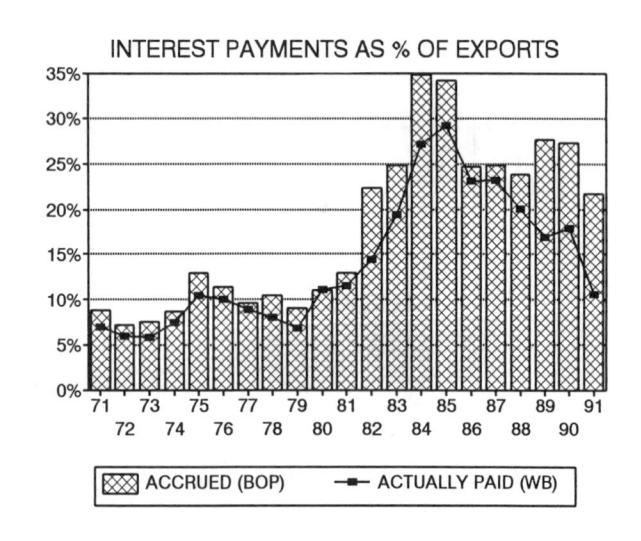

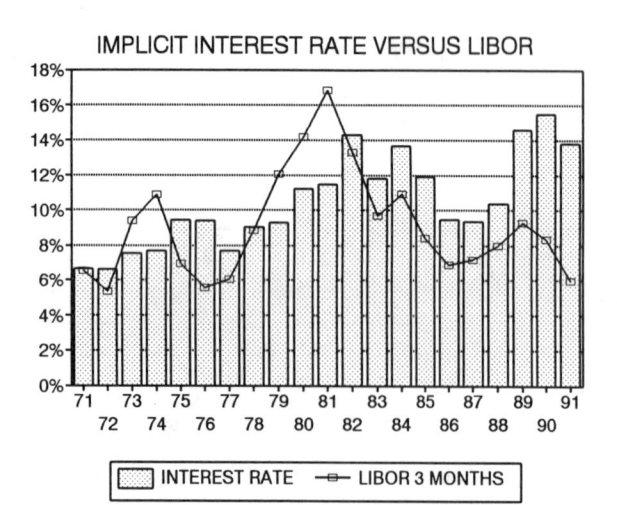

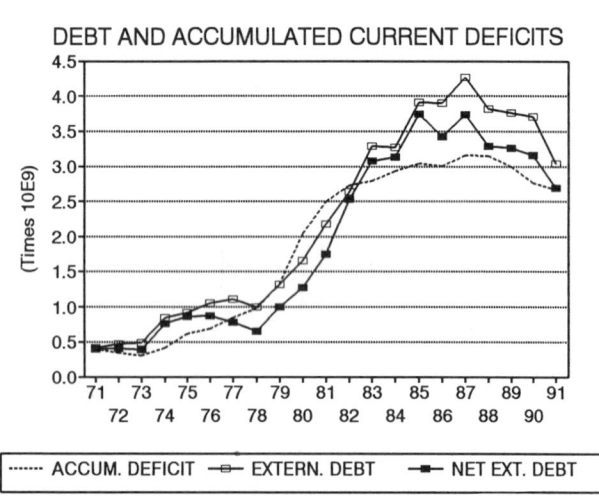

# URUGUAY
## External Debt in US Dollars

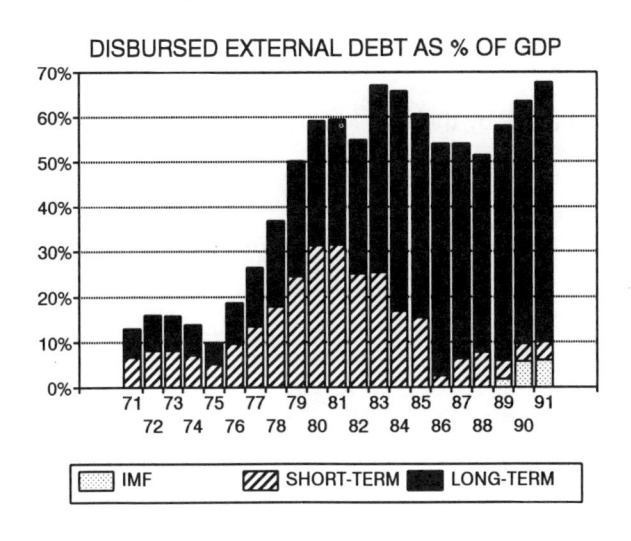

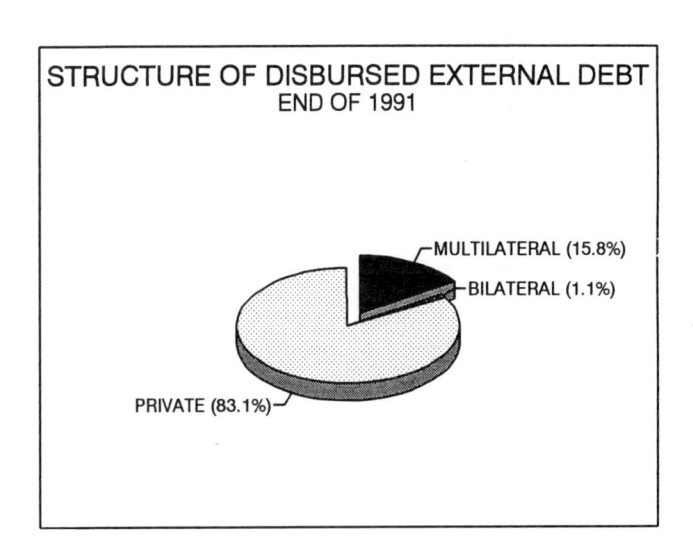

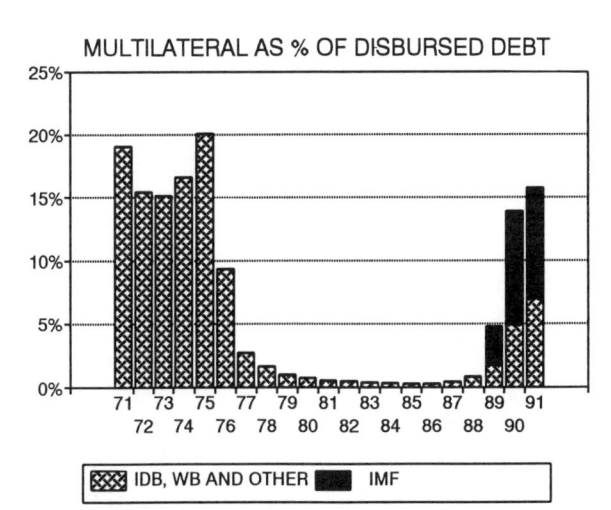

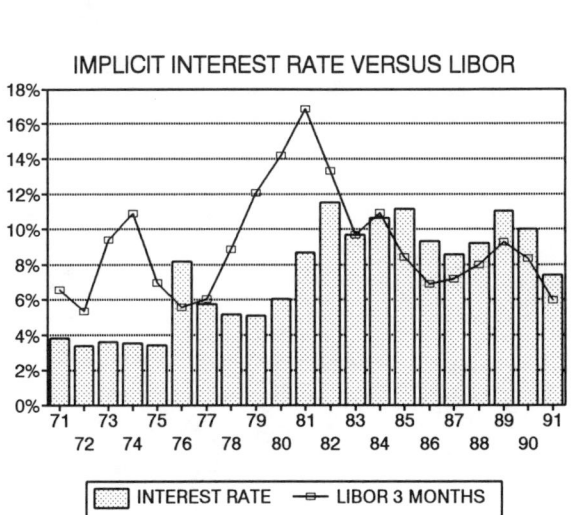

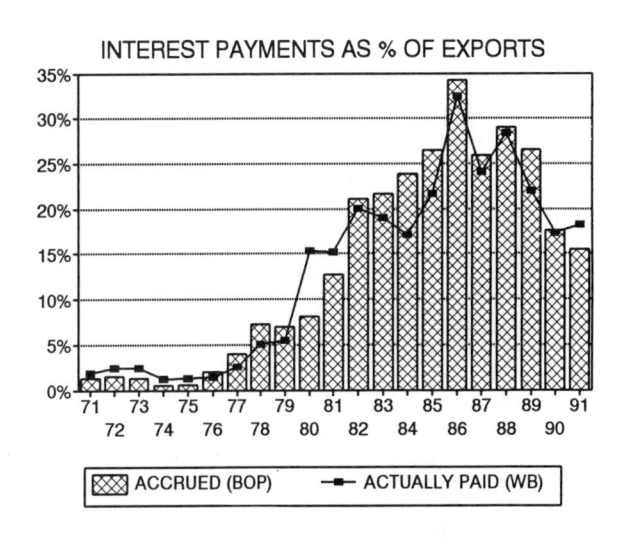

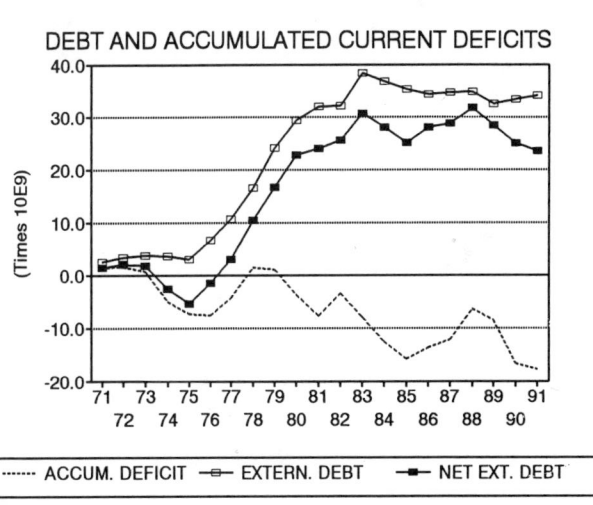

# VENEZUELA
## External Debt in US Dollars

Statistics and Quantitative Analysis/IDB

127

## Country Notes on External Debt

### Haiti
External Debt disbursed excludes the official debt write-off by the United States and France of 159.0 million Dollars in 1991.

### Honduras
External Debt disbursed excludes the official debt write-off by the United States, Switzerland and Holland of 448.4 million Dollars in 1991.

### Panama
The apparent interest rate is based on *actual* interest payments instead of *accrued* interest payments.